UNDERSTANDING DISABILITY
LAW THROUGH GEOC

To my Parents:
Mohamed and Gulshan Vellani

Understanding Disability Discrimination Law through Geography

FAYYAZ VELLANI

Routledge
Taylor & Francis Group

LONDON AND NEW YORK

First published 2013 by Ashgate Publishing

2 Park Square, Milton Park, Abingdon, Oxon OX14 4RN
711 Third Avenue, New York, NY 10017, USA

Routledge is an imprint of the Taylor & Francis Group, an informa business

First issued in paperback 2016

British Library Cataloguing in Publication Data
Vellani, Fayyaz.
  Understanding disability discrimination law through geography.
  1. Great Britain. Disability Discrimination Act 1995. 2. People with disabilities–Legal status, laws, etc.– Great Britain.
  I. Title
  342.4'1087-dc23

Library of Congress Cataloging-in-Publication Data
Vellani, Fayyaz.
  Understanding disability discrimination law through geography / by Fayyaz Vellani.
    p. cm.
  Includes bibliographical references and index.
  ISBN 978-1-4094-2806-0 (hardback : alk. paper)
1. People with disabilities–Education (Higher)–Law and legislation–Great Britain.
2. Discrimination against people with disabilities–Law and legislation–Great Britain.
3. People with disabilities–Legal status, laws, etc.–Great Britain. 4. Great Britain. Disability Discrimination Act 1995. 5. Great Britain. Equality Act 2010. 6. Law and geography–Great Britain. 7. People with disabilities–Legal status, laws, etc.–United States. 8. People with disabilities–Legal status, laws, etc.–Australia. I. Title.
  KD737.V43 2013
  344'.0798–dc23

                                                            2012030424

ISBN 978-1-4094-2806-0 (hbk)
ISBN 978-1-138-25276-9 (pbk)

Contents

List of Figures and Tables

Figures

Tables

Introduction

While this book examines disability discrimination law as a particular form of legislation which attempts to redress discrimination against disabled people, it is situated within a larger discursive context which seeks to understand the role of law and geography in shaping social and material realities. In examining the United Kingdom's Disability Discrimination Act (DDA) 1995, Equality Act (2010), the Americans with Disabilities Act (ADA) 1990 and the Australian Disability Discrimination Act (DDA) 1994, the book seeks to made wider arguments about the efficacy of law in relation to its enforcement in varied geographical contexts. In so doing, law itself is interrogated both epistemologically and ontologically.

The study conducted for the book encompassed a variety of methods – including interviews, surveys and documentary analysis – across a range of contexts both within the UK and outside it. As the various parts of the UK DDA came into effect over a number of years, the book is designed to analyse these parts chronologically and with increasing depth for the more recent parts. For example, Part 3 of the DDA concerned Service Provision and this is the focus of Chapter 3 of the book. Part 4 of the DDA, also known as the Special Educational Needs and Disabilities Act (SENDA) came into effect much later than Part 3 and is considered in great detail in Chapters 4, 5 and 6. These latter chapters devote significant attention to how Higher Education Institutions (HEI) in England and Wales responded to SENDA for two main reasons. The first is that HEIs are effective examplars as their activities encompass a wide range of the law's constituent parts; not only Part 4, but also Parts 2 (employment), 3 (service provision) and 5 (transport). Therefore, examining HEIs enables a deeper understanding of how this legislation translates across various social and geographical contexts.

The second reason for the focus on higher education is that unlike other Parts of the DDA which largely required changes to physical infrastructure or processes, SENDA required institutions to rethink the way in which education was conceptualised. Creating inclusive education impacts not only classrooms and buildings but also notions of discourse and knowledge production, and the examination of power structures which influence these discourses. In this sense, the DDA had the potential to fundamentally alter the ways in which HEIs operated and to broaden notion of what 'access' and 'inclusion' meant beyond wheelchair ramps and lifts.

In examining the higher education context, the book – in Chapter 4 – starts at the geographical scale of the nation-state and, employing the findings of a postal survey of all HEIs in England and Wales, considers the impact of SENDA on these institutions. In Chapter 5 it takes the analysis to the institutional scale and considers how two very different HEIs understood and actualised this legislation. These institutions were selected precisely for their very different profiles – one being highly selective and the other being highly inclusive – and in keeping with the importance of context in geographical research. Particular arguments about law's efficacy can be made in relation to the way it unfolds in these two different geographical contexts, highlighting law's contingency and challenging its neutrality and abstraction from social and geographical realities. The final chapter dealing with higher education – Chapter 6 – examines the law at an even more nuanced scale, cataloguing the micro-social processes and experiences of disabled students in these two institutions. This allows an examination of the relationships between legal discourses across the various scales employed, e.g. how legal definitions of disability are mapped onto space and place and woven into disabled students' life stories.

Throughout the book, several themes emerge and reappear in various chapters. The first is the finding that Western legal systems are rooted in liberal conceptions of human and social life. This is explored in Chapter 1 but appears in virtually every chapter which follows. The underlying features of liberal legalism are its framing of societies through the lens of the individual and its conceptions of citizens as free and independent social and economic agents. When applied to the context of disabled people grappling with antidiscrimination legislation, interesting observations emerge which challenge conventional liberal legal thought.

Another theme which appears throughout the book has to do with the medicalisation of disability. Foucault (1979) traces this back to eighteenth-century France and others (Stiker, 2000, for example), trace it back even earlier. Disabled people asserting their legal rights in the current context are still subject to various fora through which they must 'prove' their status as disabled and that they are less able than able-bodied people – against whose norms they are measured. Related to this are debates which emerged within the field of Disability Studies about social versus medical models of disability and these are examined in great detail throughout the book.

Flowing from the medicalisation of disability is the seeming inescapability of the categorisation of individuals into groups/classes of people – an apparently important feature of liberal legalism. For disabled people it means not only proving that they are disabled but also categorising themselves as belonging to a particular group by dint of the specificities of their impairment, e.g. deaf, visually impaired, ambulant, paraplegic etc. Many disabled people interviewed experienced these categories as themselves discriminatory, limiting and contradicting the spirit of inclusion at which the law aims. Following on from this, disabled research subjects repeatedly called attention to their perceptions about the invisibility of able-bodied privilege. This theme also appears throughout the book and appears to be most acutely experienced (and expressed) in the context of the rarefied bastions of HEIs. In interview after interview, disabled people pointed out that the entire institution – which can be viewed as a microcosm of society – was designed and built for able bodied people and predicated on the view that variance from this unspoken assumption is an aberration to be accommodated or tolerated at best but not truly included.

A final theme which emerges is about the limits of law itself (Schuck, 2000). In highlighting the social and geographical contingencies which law fails to adequately address, it becomes apparent that law itself – as it is currently conceptualised – does not change social or geographical realities. Various theorists from Harvey to Unger have posited new ways of humans relating to each other which radically call into question existing social arrangements and which would fundamentally alter current political and legal systems. While making such a case is not within the scope of this book, it reaches the conclusion with the awareness that law itself requires deeper examination and cannot escape such scrutiny if equality and inclusion are to be achieved.

Liberal legalism – and its alliance with free market capitalism – are currently facing an existential crisis and their effectiveness at creating individual and social good are being questioned given the banking crisis of 2008 and the currency crises of 2011. At present there are also debates about the efficacy of law in relation to current events, as witnessed in the controversies around how rigidly or loosely the criminal code should be applied to those youths charged following the August 2011 riots in the UK. The debates highlight the contingency of law itself and have brought concerns about the uneven application of law into sharp focus. This book therefore finds itself looking at a specific law – the DDA – but with implications which fit into current debates about law itself, the power of legal enforcement mechanisms, and contested ideas about how resources should be allocated in times of economic scarcity. Disabled people in particular have been casualties of swingeing cuts and their experiences under the Coalition Government elected in May 2010 suggest that the legal enshrinement of their rights alone is not enough to guarantee their ability to sustain themselves independently and with dignity.

The Equality Act [2010]

The Equality Act [2010] was introduced to replace and harmonise legislation governing UK equality writ large including the Sex Discrimination Act [1975 and 1986], the Race Relations Act 1976, the Employment Act 1989, the Disability Discrimination Act (DDA) 1995 and the Equality Act [2006]. It will therefore take some time before critical legal perspectives on the Equality Act [2010] are gleaned due to the dearth of case law arising in its wake and because of the change of government in 2010.

The remaining chapters of this book engage in detail with the Disability Discrimination Act (DDA) 1995 and the Special Educational Needs and Disabilities Act (SENDA) 2001 as they set the stage for how disabled people experienced and engaged with public space – and how disability itself was legally conceived – from 1995 to 2010, a formative period for UK antidiscrimination law culminating in the Equality Act 2010. The DDA also laid the social, political and legal groundwork for how Equality Act [2010] envisages equality, disability and discrimination. In effect, the same foundational principles – of seeing disability as an aberration from 'normal day-to-day activities' and thereby medically construed from a deficiency model – and of relevant parties making 'reasonable adjustments' so as not to place disabled people at a 'substantial disadvantage' remain firmly ensconced in the Equality Act. Therefore, the remaining chapters provide an analysis of legislation which, while not extant in its current form, remains alive in the spirit of its successor, the Equality Act [2010].[1]

1   For more detail on the Equality Act 2010, please see the Epilogue.

PART I
Concepts in Disability Law

Chapter 1
Law and Liberal Legalism

(1.0) Introduction

This book investigates the role and importance of law and legal frameworks in seeking to secure accessible environments for disabled people. The research for the book was primarily conducted in England and Wales with some comparative research done in Australia and the USA, examining the responses of service providers to legal requirements to be more inclusive of disabled people. In particular, universities represent a category of service providers which is particularly interesting because access to higher education entails not only physical access, but also access to power structures which determine the experiences of disabled students. Larger questions of how socio-institutional frameworks and cultural assumptions about disability materially influence disabled people can be viewed in the context of inclusiveness at universities.

The objective of this book, then, is to explore the scope of the Disability Discrimination Act (DDA) 1995, in redressing the marginal status of disabled people in England and Wales.[1] How far is this legislation transforming, or how far will it be able to transform, disabled peoples' lives? Liberalism has dominated western legal systems and so, to answer this question I examine liberal conceptions of law and critiques of these conceptions (Blomley, 2000). The chapter is divided as follows: (1.1) The scope of law and legal discourses, (1.2) The roots of liberalism: utilitarianism, (1.3) Liberalism, law and the legal subject, (1.4) Epistemologies of law, (1.5) The scope of law in overturning discrimination, and (1.6) Space, place and specificity: a critical legal geographical approach. In (1.7) I conclude the chapter.

(1.1) The Scope of Law and Legal Discourses

Over the past forty years, there has been a considerable rise in the number and activity of new social movements and protest groups in the UK,[2] USA, Australia and elsewhere (Young, 1990). One of the goals of such groups, including women,

---

1   While this study is focused on England and Wales, I was able to conduct small-scale comparative studies in the USA and Australia, and so the book draws on materials from those studies.

2   In this chapter, I refer to the United Kingdom, as that is the national government which oversees England, Wales, Scotland and Northern Ireland, and the DDA applies to all four of these jurisdictions. In Chapter 2, I refer to the UK in comparison with the USA and

people of colour, gays and lesbians, and disabled people, has been the pursuit of legislation which seeks to overturn discrimination (Blomley et al., 2001; Gooding, 1994; Handley, 2001; Young, 1990). These groups have used the language of rights to redress their marginalisation, with much of the inspiration arising in the wake of the American civil rights movement (Gooding, 1994). However, as many argue, this approach alone, without other measures to counteract cultural assumptions and norms in everyday life, for example, will not suffice in improving the lives of disabled people (Handley, 2001; Young, 1990). Critical legal theorists, feminists, anti-racists and disabled scholar argue that the law – in its current Western, liberal forms – has limited capacity to change people's lives (Clark, 2001; Razack, 1991; hooks, 2003). The extent and scope of these capacities are scrutinised in this chapter.

These arguments are manifold, but rest primarily on one question. While the law itself is a complex tool which shapes and governs peoples' lives, can legislation reach areas of human interaction such as cultural biases, prejudices and norms, as well as the broader framework of values in western liberal democracies (Gooding, 1994; Young, 1990)?[3] For example, disabled activists have lobbied for, and have been successful in obtaining, CRL in European and North American countries (Campbell and Oliver, 1996). Some theorists argue that this legislation alone is not enough to change disabled peoples' lives, as it cannot transform wider systemic disadvantages faced by them (Oliver and Barnes, 1998). For example, Gooding (1994), Oliver and Barnes (1998), and Stiker (2000) argue that antidiscrimination legislation (ADL) does not fundamentally call into question, or seek to alter, wider socio-legal and economic systems in which peoples' value to society – in basic terms – is primarily measured through economic output and productivity.[4] Others question whether ADL will be able to permeate into people's everyday interactions with each other, where subtle forms of prejudice are encountered (Gooding, 1994; Young, 1990).

There have also been critiques of the ontological basis for ADL, beyond the limitations of its liberal legal values. Specifically, the DDA is premised on the medical model of disability, rather than the social one, and as demonstrated in later chapters, the use of the medical model has material consequences for disabled people (Gooding, 1994; Oliver and Barnes, 1998; Young, 1990). Moreover, the DDA is premised on the idea that disability is primarily an individual phenomenon, rooted in the impairment of specific individuals. Conversely, in a social model of disability, society is seen to disable those with impairments by building barriers impeding their full access to social life, thus removing the stigma from the individual (Barnes et al., 2002; Oliver and Barnes, 1998).

Australia. In Chapters 4, 5 and 6 I refer to England and Wales, as my empirical study of HE deals only with those two jurisdictions.

3   For a discussion of values, see Chapter 6.

4   For more detailed arguments on how the DDA does not change existing economic arrangements, see Chapter 3.

In the UK DDA, and its Australian counterpart, also called the DDA, for example, the onus is on individual disabled people to pursue their legal cases where they feel they have been discriminated against, rather than a proactive approach in which discrimination is broadly discouraged at the societal level[5] (Handley, 2001; Roulstone, 2003). A proactive approach, for example, would aim to influence the behaviour of organisations and institutions so that they are more inclusive of disabled people, rather than simply attempting to accommodate disabled people within existing social structures (Barnes et al., 1999). Such an approach would more closely resemble civil rights legislation (CRL) such as the Americans with Disabilities Act (ADA) but this law too, has its limitations, as will become clear later in the book. Some theorists would prefer a legislative approach where the law seeks to discourage discrimination before it happens, rather than after the fact, a more overt goal of CRL than of ADL (Barnes et al., 2002; Barnes et al., 1999).

CRL, at the very least, has been symbolically significant for disabled people and has established a legal framework for a more egalitarian distribution of opportunities and resources within the existing socio-economic system (Gooding, 1994; Oliver and Barnes, 1998; Young, 1990). For example, the Americans with Disabilities Act (ADA) has been successful in increasing access to the built environment in the United States of America (USA).[6] There has also been more employment of disabled people in the USA since this legislation was enacted in 1990 (Gooding, 1994).

These may be seen as either positive or negative developments, depending on one's point of view (Bellamy, 2000; Ramsay, 1997). Some argue that the ADA has not called into question able-bodied peoples' fundamental attitudes about disability, and that it does not recognise the intrinsic value of disabled people, but rather makes them adapt to able-bodied labour markets and built environments (Gewirtz, 1997; Hahn, 2001). Others argue that law alone cannot change fundamental attitudes, and that much more is required for these attitudinal shifts (Young, 1990). This chapter evaluates the efficacy of ADL and CRL in changing disabled people's lives, and does not necessarily to draw conclusions about the wider socio-political projects within which they are embedded. In order to evaluate this legislation, and unearth its values and assumptions, I turn to the origins of western liberal philosophies: utilitarianism.

---

5   The Americans with Disabilities Act (ADA) allows the US government's Department of Justice to be more proactive in prosecuting disability discrimination than its UK or Australian counterparts. For a discussion of all three laws in their national contexts, see Chapter 2.

6   See Chapter 2 for more comprehensive analysis of the ADA.

(1.2) The Roots of Liberalism: Utilitarianism

In developing critiques of liberalism, it is important to examine the origins of its current manifestations. Contemporary legal thought has been significantly influenced by utilitarianism (Clark, 2001; Pue, 1990). Utilitarianism originated in England as a philosophical inquiry into the workings of government and the feudal economic theories which existed prior to the industrial revolution of 1750 to 1830 (Dent, 1993; Mill, 1859). Utilitarians advocated a series of political, social, and economic reforms based on the principles of liberalism and the need for more efficient and responsive government (Fitzpatrick and Hunt, 1987; Handley, 2001; Ward, 1998).

Utilitarianism originated as the study of law based on simple rules and assumptions about human nature and the motivations behind human behaviour (Dent, 1993). Mill (1859), in his book *On Liberty* – the classic utilitarian treatise – applied post-Enlightenment scientific and rational methods to ethics and politics. The philosophy of utilitarianism was based on the principle of utility, or the ability to please as many individuals as possible, thereby achieving the greatest good for the greatest number of people (Mill, 1859). As Dent (1993:7) notes:

> Utility, or the Greatest Happiness Principle, holds that actions are right in proportion as they tend to promote happiness, wrong as they tend to produce the reverse of happiness. By happiness is intended pleasure, and the absence of pain; by unhappiness, pain, and the privation of pleasure... pleasure, and freedom from pain, are the only things desirable as ends.

Utilitarians theorised that human behaviour is guided by the desire to avoid as much pain and seek as much pleasure as possible (Dent, 1993; Handley, 2001; Ward, 1998). The utility of any action, therefore, would depend upon the minimisation of pain and maximisation of pleasure resulting from it (ibid.). The role of government, in this view, is to achieve utility using a simple pleasure-pain scale to evaluate actions, with the most favourable actions giving the most pleasure to the largest number of individuals (Dent, 1993). Utilitarian reforms had significant impacts on social policy and practice from the mid 1800s onwards – including better conditions for the working classes in England and the United States – and utilitarianism came to be viewed as a pillar of democracy (Fitzpatrick and Hunt, 1998).

Contemporary western democracies have legal systems premised on utilitarian notions that law, by improving the lives of individuals, improves the whole of society (Birch, 1998; Dearlove and Saunders, 2000). For example, in the socio-economic realm, citizens who self-actualise to their highest economic potential accrue benefits personally through wealth creation, and pass on these benefits through taxation, and via consumer spending, which creates jobs and opportunities for others (Young, 1990). Therefore, one of the predominant aims of human life in contemporary liberal capitalist contexts is economic self-actualisation (Stiker, 2000; Ward, 1998).

In liberal economic theory, all that is needed to create maximum utility for all individuals is a level playing field, an oft-used metaphor for the competitive nature of capitalist market economies (Rose, 1987; Young, 1990). However, the presumption of a level playing field ignores broader systemic inequalities which may impede certain groups, such as disabled people, from equally participating in social and economic life (Gooding, 1994; Razack, 1991; Young, 1990). The existence of a playing field, i.e. the labour market, does not necessarily signify equality or a society which is truly inclusive (Fitzpatrick and Hunt, 1987; Gleeson, 1999). For example, Gleeson (1999:38) describes Marx's critiques of capitalist exploitation and neglect of disabled people as follows:

> For Marx, the suffering 'pauper' – that spectral other which haunted the Victorian bourgeois imagination – was testimony to capitalism's oppression and exploitation of the physically vulnerable. His pauper was a polymorph whose many forms included vagabonds, criminals, and prostitutes – the 'lumpenproletariat' – together with those in the proletariat who had failed in the competition to sell labour powers. For the latter, Marx clearly recognised physical infirmity as a principal cause of unsaleable labour power.

Given this history of exclusion of disabled people from labour markets, many argue that liberalism's focus on self-fulfilment via individual choice ignores social attitudes and structures which may impede disadvantaged members of society from asserting their agency (Gooding, 1994; Young 1990). Some critical legal scholars, such as Blomley and Clark (1990:13), claim that "the individualist obsession of legal liberalism offers only a "thin" – even barren – model of social life." At the same time, they warn against a wholesale dismissal of liberalism, noting that some of its values, such as the legal promise of equality and justice, are laudable (ibid.). As Blomley and Clark (1990:14) note:

> It is vital at this point, however, to note that scepticism toward liberalism need not imply a simple-minded rejection. Although the legal critique can appear irreverent, it is also clear that certain principles of legal liberalism are regarded as viable and as normatively valuable. Thus, the legal promise of equality and justice – albeit partial – contained within the concept of the rule of law," has been applauded by some radicals ... similarly, the appeal of liberalism to tolerance, autonomy..is undoubtedly appealing.

As noted earlier, this book and chapter do not provide polemical accounts of the DDA, but instead seek to understand how disabled people experience law and legal processes. The next section highlights how those experiences have been ensconced in particular liberal conceptions of human beings in relation to each other, and the law.

(1.3) Liberalism, Law and the Legal Subject

Critical legal theorists and writers from other disciplines have argued that liberal values limit the ability of the law to redress social inequality because in liberalism, law is construed as being abstract, neutral and removed from society (Blomley et al., 2001; Clark, 2001; Pue, 1990; Ramsay, 1997; Razack, 1998; Young, 1990). Moreover, because law is defined as an autonomous field of inquiry, liberals see it as being 'objective' and sufficient unto itself, as something which stands abstractly apart from the 'subjective' and 'particular' nature of society (Clark, 2001; Pue, 1990). This view can limit the ability of law to redress social inequality because in societies which exhibit systematic inequality, an insistence on the rule of law can serve to legitimate those inequalities (Fitzpatrick and Hunt, 1987; Minow, 1991; Razack, 1998; Razack, 1991). For example, feminists have argued that while law claims a neutral stance, it has traditionally favoured men, because the liberal legal subject has historically been conceived of as male (Ramsay, 1997). The next section examines various terms and concepts associated with law in liberal societies and consider what different theorists have to say about them.

(1.4) Epistemologies of Law

Four paradigms through which legal theorists view the law are positivism, realism, instrumentalism, and legal formalism. Positivism assumes institutional structures as given rather than bringing them under normative evaluation (Young, 1990). For example, a positivist view of discrimination against disabled people would remedy the problem by making discrimination illegal (Razack, 1998). When positivism is applied to law, law is assumed to be capable of responding to society's needs and solving its problems (Gooding, 1994). Critics argue that a positivist view of law ignores larger social structures, and/or other factors which are not directly within the realm or scope of law, or cannot be addressed in such straightforward terms (Blomley, 1994; Minow, 1991; Ramsay, 1997).

Blomley (2000), Clark (2001) and Roulstone (2003) argue that a positivist approach to the law would imply that the law is straightforward and 'natural'.[7] In Pue's (1990:570) words, for lawyers, positivism is the view that "the law is not mysterious to those who are trained to look for it." Realism, on the other hand, is a concern for fact or reality and a rejection of the impractical and visionary (Clark, 2001). Altman (1990:152) calls a realist view of law "rule scepticism." Legal realism posits that the law, in asserting itself as neutral and unbiased, is legislated, enacted and enforced by people, who, regardless of their attempts at justice, will deliver and interpret these laws in varying ways, depending upon their own understandings (Clark, 2001). A realist view of law thus rejects law's positivist claims to value-neutrality as it suggests that the law only exists as interpreted

---

7   Roulstone (2003), for example, criticises naturalist conceptions of law.

through the people in charge of it, and can claim no moral absolution or justice which is completely consistent in its treatment of all legal subjects (Pue, 1990). A realist view of the law would acknowledge, for example, that the personal biases of judges do affect the legal decisions in court cases (Bagguley, 1990; Pue, 1990).

Instrumentalism, according to Blomley (2000:10), is the view that law is "an autonomous instrument with which an activist legislator can bring about (or prevent) social and economic change." An instrumentalist view of law assumes that 'society,' as an entity separate from 'the law' has various needs, and that 'the law' responds to these needs and 'acts upon' society to achieve the desired results (Blomley, 2000). According to Blomley (2000), this view ignores the organic nature of the relationships between law and society. For example, an instrumentalist approach would downplay the fact that law is a product of a historical evolution involving complex relations between societal actors and lawmakers. Instrumentalism assumes that society is separate from law, and can be acted upon by law (Clark, 2001). Although there are some similarities between instrumentalism and positivism, in the latter the law is seen as straightforward and self-evident (ibid.). Moreover, the main aim of legal positivism is to restrict legal theory to description rather than prescription (Ward, 1998). Instrumentalism, on the other hand, sees law as being able to prescribe solutions to society's problems (Blomley and Clark, 1990).

Legal formalism represents the closure of law to outside influences (Blomley, 2000). In a formalist view of law, legal rules form a consistent, complete whole from which the answer to any legal question can be logically deduced by discovering the applicable rule and applying it to the facts of the case (Altman, 1990). Law and politics are seen to be separated, and legal rules should therefore not be changed (ibid.). These rules are seen to be 'closed' to political or other influences from 'outside' the law, thus legitimating their authority as independent, neutral and just (Clark, 2001). In this view, the facts of law are seen to be objective and self-sufficient (Dworkin, in Blomley and Clark, 1990). In fact, Blomley (2000) says that the closure of law is *central* to western law and legal practice, and I discuss legal closure further in Section (1.6) in relation to critical legal geographical theories.

In legal formalism, the law is characterised as autonomous, self-sufficient, and therefore removed "from the vagaries of social and political life" (Blomley, 2000:7). For example, Hart (2001) points out that in western liberal societies, law is often seen through the eyes of judges and lawyers (in Clark, 2001). Hart (2001) notes that there is no legal theory extant which has adopted a victim or defendant perspective (in Clark, 2001). Furthermore, he says that liberal legalism takes the self-descriptions of judges and lawyers as primary empirical material (ibid.). In this view, judges' stated views on what they do and why they do it are treated as direct evidence about the nature of legal practices (Fitzpatrick and Hunt, 1987; Razack, 1998). The difference between realist and formalist critiques of law seem ambiguous in this example, which illustrates the lack of clearly defined boundaries in the field of critical legal studies. While these theories are fluid and contestable, they all seek to question the supremacy of the rule of law in liberal societies.

Critiques of formalism centre on the view that law is implicated in the politics of social life and is *itself* constitutive of social and political relations (Blomley et al., 2001). Critical legal theorists see critiques of formalism as a kind of 'opening' of the law (Blomley, 2000).

Formalism is a key function of liberal bureaucracies (Young, 1990). As Young (1990:78) notes, "the formalism, universality and impersonality of the rules are supposed to protect persons from the arbitrariness of whim and personal likes and dislikes – everyone is to be treated in the same way, impersonally and impartially, and no particular values should enter." However, she points out that the people applying these impersonal rules must make judgements about how they apply to each case, and in this application, the feelings, values and perceptions of decision-makers inevitably enter (Young, 1990).

Judicial decision-making, for example, is seen from a formalist perspective as being constructed to appear neutral, objective and universal (ibid.). In critiquing conceptions of the law as rational and neutral, Clark (2001) makes two propositions regarding the example of judicial decision-making. His first proposition is that judicial opinions are like narrative texts "in that they aim to justify certain arguments and/or opinions in contradistinction to competing representations of the facts of a case" (Clark, 2001:107). His second proposition is that legal reasoning "is not like literary text-based reasoning in one obvious and profound way: the majority opinion is the immediate and authoritative interpretation of a case" (ibid.).

Given these propositions, Clark (2001) says that judicial decision-making is neither as democratic nor as tolerant as literary interpretative practice. Stemming from these propositions, he argues that courts have a powerful role in making determinate conclusions about the proper structure of society, and that their power is manifested and perpetuated in the following three ways: (1) idealising the law as an institution, (2) idealising the nature of legal reasoning as a form of discourse, and (3) idealising the normative image it authorises (ibid.). This is problematic for disabled people, women, and other individuals who do not conform to idealised legal norms. As I demonstrate in Chapter 2, law can not only exclude disabled people, but through the perpetuation of able-bodied bias in courtroom proceedings, law may cause harm and damage to disabled people's confidence and sense of well being (Roulstone, 2003).

Indeed, critical legal theorists, as well as others, contest the notion of impartiality (Blomley et al., 2001; Clark, 2001, Gooding, 1990; hooks, 1982; Minow, 2001; Ramsay, 1997; Razack, 1998, Young, 1990). For example, Young (1990) argues that the ideal of impartiality expresses a view of identity which seeks to reduce differences among people and situations to unity. She posits that notions of detachment in justice systems stem from an abstraction from social realities (ibid.). Young finds this problematic in two ways. First, she argues that the particularities of individuals and contexts cannot be removed from moral arguments (ibid.). Second, she says that law's claim to impartiality hides the ways in which dominant groups' perspectives claim universality and justify decision-making structures which are hierarchical (Young, 1990).

Feminist theorists argue that law, in western liberal democracies, has largely been shaped by white, able-bodied, heterosexual men (Gooding, 1994; hooks, 2003; hooks, 1994; hooks, 1982; Minow, 1991; Morris, 1991; Razack, 1998; Razack, 1991; Young, 1990). It has traditionally been assumed that legal subjects have the same gender, race, ability and sexual orientation, and this image has become projected as a universal one (ibid.). As discussed earlier, this is an example of reductionism. By evaluating or treating all individuals based on laws which have historically been written by white able-bodied heterosexual men, the western liberal application of law reduces the legal subject to the narrowest possible definition (Razack, 1991). Difference and individuality are wiped out; the law only sees the image it projects onto its subjects (Young, 1990).

In seeking to treat all legal subjects from a universal, impartial, objective stance, the law seeks to reduce difference into sameness (Young, 1990). The result is that the variant becomes the 'other' (Foucault, 1991; Said, 1978). The more different people are from 'universal norms', which are not really universal, the further outside of the political and legal mainstream they are positioned (Young, 1990). Gooding (1994) argues that in order to assess the law's powers and limitations, we should critically examine the law's claims to impartiality. As she says (1994:28):

> Both feminists and anti-racists have asserted the partiality of legal discourse, not only despite but because of its claims to universality and rationality. The law has historically only been capable of putting forward its partial viewpoint as a universalist one by discounting whole categories of persons – women, slaves, aliens, and disabled people …

Gooding (1994) alludes to the fact that there are many who oppose liberal legal paradigms and contest the law's claims to rationality, normative worth, and efficacy. Such challenges have come from diverse arenas, including feminists, Marxists, and, particularly, critical legal theorists (Blomley and Clark, 1990). In critical legal terms, law is relational and acquires meaning through social action. Law is socially constructed and thus inseparable from social and political relations (Blomley and Clark, 1990). Not only does law structure the manner in which we experience social life, and vice versa, but legal discourses construct our social roles and shape the definitions of who we are, for example in relation to occupational and marital status. In Blomley and Clark's view (1994:12):

> Legal discourses, it is claimed, split the world into categories that filter our experience, distinguishing a set of harms that we must accept as the hand of fate or our own fault – such as poverty – from those actions that we may legitimately contest – such as libel or assault in a public place.

Disabled people, for example, are categorised and pigeonholed by law (Morris, 1991). The kinds of opportunities available to disabled people, in terms of housing, transportation, education, for example, are dependent on meeting criteria for certain categories and severities of impairment (Oliver and Barnes, 1998). In this way, the law does directly constitute disabled people's social reality.[8] I will discuss law's role in constituting social and material space in Section (1.6).

Beyond constituting social reality, law is seen as deeply social in terms of a pervasive legal mentality which Foucault terms governmentality[9] (in Dreyfus and Rabinow, 1983). Law is seen as an idealised social vision. In Western societies this vision belongs to a particular set of liberal values. As Blomley and Clark (1990) argue, liberal ideas profoundly influence law's 'social vision.'

The reliance of Western legal systems on liberal values might be problematic for disabled people in that liberalism is primarily concerned with individual rights (Ramsay, 1997). There are debates about the relative merits of legal systems focused on individuals rather than on social or communitarian perspectives. Those who favour liberal legal approaches to transforming disabled peoples' lives argue that the individual rights accorded through CRL, a Bill of Rights, or a Constitution, for example, are empowering and ennobling (Amar, 2000). Rather than being categorised as belonging to a group through which these rights are accorded, disabled people, like all other members of society, are equal and free individuals, as recognised and reinforced by CRL.

An opposite view posits that liberalism stems from the utilitarian view that all individuals should be allowed to do as they please, so long as their actions make them happy (Dent, 1993). Regardless of the relative theoretical merits of individual liberty, critics point out that without equality of opportunity, legal equality is powerless (Fitzpatrick and Hunt, 1987; Ward, 1998). Phillips (2004), for example, notes that pursuing equalities of outcome may be more equitable than merely providing equal rights. Throughout this book I interrogate the boundaries of legal rights in terms of their efficacy.

In this section I have analysed epistemologies of liberal law – positivism, realism, instrumentalism and formalism – as well as critiques of these foundations, arising from diverse sources such as critical legal scholars, feminists, lesbians and gay people, ethnic minorities, and disabled people. I have demonstrated how contemporary views of law in Western liberal democracies are informed by utilitarian principles, and how an able-bodied white male heterosexual perspective has been projected onto law as the universal legal subject. Finally, I have examined debates around the efficacy of liberal concepts in creating equality; indeed, some argue that liberalism should not be discounted altogether. As Gooding (1994:30) notes, regardless of the shortcomings of liberal rights-based laws,

---

8   For example, in Chapter 6, I examine how the category of 'special needs' had material impacts on disabled students' educational experiences.

9   Governmentality can be defined as a set of practices, norms and beliefs and thus a technology of governance which pervades thought and action from the highest to the lowest levels of bureaucracies (Foucault, in Dreyfus and Rabinow, 1983).

the existence of a broader rights-based framework can prove a potent weapon for demanding more resources, and powerfully influence the context in which resources are delivered.

This analysis of the value of resource distribution in overturning discrimination will be examined in the next Section, (1.5), as it is a major source of contention for Young (1990); Section (1.6), engages with critical legal geographies.

(1.5) The Scope of Law in Overturning Discrimination

Discrimination is defined as (1) the process by which two stimuli differing in some aspect are responded to differently, (2) the act, practice, or an instance of discriminating categorically rather than individually, (3) prejudiced or prejudicial outlook, action, or treatment[10] (Merriam-Webster, 2002). Discrimination in current legal discourse means events, items, or outcomes in which the person discriminated against is treated 'less favourably' (DRC, 2001). In legal terms, the discriminatory action taken must be measurable and provable in order to have a claim brought against it. This, of course, leaves grey areas which are not easily measured, but for now this definition of discrimination serves to provide background to what is meant, in legal terms, by discrimination.

In contemporary Western societies, law has an overarching function to maintain order among society's various actors. As Blomley et al. (2001:xiv) note, law's reach is vast:

> It seems uncontroversial, even banal, to say that so much of the world we live in is shaped by and understood (by ordinary people as well as experts) in terms of law. Our everyday conceptions of authority, obligation, justice and rights, our dealings with others and our relations to collective institutions such as the state are all structured, in part, by legal norms, discourses and practices.

Many theorists agree that law can be seen as constitutive of social reality, social relations, the institutional world, the nation-state, the market and the family (Blomley and Clark, 1990; Pue; 1990). A critical legal perspective seeks to counter the prevailing view in Western liberal democracies that law is a given, and an underpinning of socio-institutional actions rather than constitutive of them (Imrie and Thomas, 1997).

As demonstrated in (1.4), current systems of law in Western countries are premised upon liberal values. These values influence the shape of ADL and CRL and the outcomes of these laws for disabled people. In a liberal view, law is rational, ordered and closed to socio-political influences and is construed as

---

10   I suggest that condition (2) of this dictionary definition is problematic and indicates the kind of tautological thinking behind definitions of discrimination.

abstract and acontextual (Clark, 2001). As Pue (1990:570) notes, law is viewed, in legal scholarship, as "a thing which stands apart from human society."

As noted earlier, impartiality is a highly contested notion (Blomley et al., 2001; Clark, 2001; Young, 1990). Laws are developed by humans and may be influenced by the humans who produce them, through social, emotional, personal and political influences. As Blomley et al. (2001) note, the specific social contexts within which law is applied are a conditioning aspect of those laws, and of the legal processes which stem from those contexts. In the case of disabled people, if all or most of those developing CRL for disabled people are able-bodied, theorists such as Young (1990) think it is worth considering whether their own biases could affect the form and content of those laws. This is in tandem with a broader literature on the nature of the state, its subjectivity as an entity, and its claimed neutrality/ universality (Foucault, 1979). For example, from a feminist perspective, the state embodies partial interests (Ramsay, 1997).

As Ford (2001) argues, no political system is completely neutral from the influence of social and geographical contingencies, and no system is without bias (in Blomley et al., 2001). Therefore, not only do ADL and CRL need to help overturn discrimination against disabled people but – to be effective – they may also have to counter the legal system's own potential biases against those outside of the legal mainstream (Gooding, 1994). For example, Schmitt's (1976) concept of the universal rule of law can be contested on the grounds that the law has historically been exclusive of disabled people, and inaccessible to them. Some suggest that the law, as a tool for disabled people, should be useable, accessible, empowering and democratic (Barnes et al., 1999). This can lead to an inside/ outside distinction in which the law is autonomous, and legal subjects are on the outside, unable to question or influence the inside content of the law (Fitzpatrick and Hunt, 1987). Moreover, as Gooding (1994) notes, merely giving people rights in law is not enough in Western liberal democracies because it assumes a level playing field for all members of society.

There are also debates about in western liberal societies about the materials impacts of rights and about how they can be equally distributed. Bellamy (2000) argues that Western political discourses are largely dominated by rights, but that we seldom examine what those rights are or what they mean. Young (1990) proposes that the language and impacts of rights must be examined in the context of what she calls *The Distributive Paradigm of Justice*. In this paradigm, in relation to contemporary western states, Young (1990) defines justice as something which is measurable in terms of the delivery of goods and opportunities to members of society. She argues that this resource-based approach in liberal states has narrowed the focus of justice to a tangible, measurable 'thing' without considering other injustices and inequalities which are not necessarily quantifiable (ibid.). For example, in allocating more resources and rights to women, some feminists argue that broader social structures which discriminate against women, such as patriarchy and the family structure, are ignored or unchallenged (Ramsay, 1997). Young (1990) feels that as long as governments continue to define justice within this distributive paradigm, the power of law and social policy are limited in their ability to redress inequality.

ADL and CRL need to go beyond the distributive paradigm, because, as Young (1990:25) points out, this paradigm limits a notion of justice which forces society to fundamentally alters its ways:

> This entails applying a logic of distribution to social goods which are not material things or measurable quantities. Applying a logic of distribution to such goods produces a misleading conception of the issues of justice involved. It reifies aspects of social life that are better understood as a function of rules and relations than as things.

Arendt's (1998) conception of participatory, open, democratic justice, for example, is broader than distributive notions of justice. Arendt (1998) believed that justice should belong to everyone. In examining how ADL and CRL can distribute rights to disabled people, there are two main areas of critique. The first is that these rights are the same rights already accorded to all non-disabled people and thus do not promise disabled people anything to which they are not already entitled (Morris, 1991). A second critique is that rights are not measurable goods, and as such when given to people by law, do not in and of themselves guarantee equality or justice (Young, 1990). As Young (1990:25) points out:

> Rights are not fruitfully conceived as possessions. Rights are relationships, not things; they are institutionally defined rules specifying what people can do in relation to one another. Rights refer to doing more than having, to social relationships that enable or constrain action.

Likewise, Foucault (1979) notes that power is not absolute, but relational and exercised in relation to and with others. As I detail in Section (1.6), Foucault's (1979) theories of power can be useful in obtaining geographical understandings of law's role in producing space. Throughout this thesis, and especially in Chapters 4, 5, and 6, I engage with Foucauldian theories of power and his critiques of institutions to examine liberal discourses of rights and equality.

In acknowledging the limits and biases of Western liberal legalism, it is important to consider how the DDA will redress the marginalisation of disabled people, when this law is a product of the legal systems which have contributed to this marginalisation (Oliver and Barnes, 1998). Young (1990:54) notes that:

> Liberalism has traditionally asserted the right of all rational autonomous agents to equal citizenship. Early bourgeois liberalism explicitly excluded from citizenship all those whose reason was questionable or not fully developed, and all those not independent ... Thus poor people, women, the mad and the feebleminded, and children were explicitly excluded from citizenship ... (Pateman, in Young, 1990).

Young (1990) makes the argument that in contemporary societies, the exclusion of 'dependent' people from equal citizenship rights is hidden just beneath the surface of social consciousness. She claims that because of their dependence on bureaucratic institutions for support and services, some disabled people receive patronising and demeaning treatment (1990:54):

> Being a dependent in our society implies being legitimately subject to often arbitrary and invasive authority of social service providers and other public and private administrators, who enforce rules with which the marginal must comply, and otherwise exercise power over the conditions of their lives. In meeting needs of the marginalized, often with the aid of social scientific disciplines, welfare agencies also construct the needs themselves. Medical and social service professionals know what is good for those they serve, and the marginals [sic] and dependents themselves do not have the right to claim to know what is good for them.

Dependency in contemporary western societies, according to Young (1990), implies permission for the government to limit disabled people's rights to privacy, respect and individual choice. This may be one aim of ADL and CRL: to give disabled people rights to privacy, respect and individual choice.[11] Indeed, the notion that all we can do to help individuals in our society is to allow them to make their own life choices is a very liberal one. Having the privacy to make one's own choices is paramount in this view (Ackerman, 1998, in Ward, 1998). How will the DDA do this, and, particularly, how will the law mandate that disabled people be respected? Respect is not a quantifiable commodity; it is given to us by others (Young, 1990).

How can privacy and respect be maintained, when, under the DDA, disabled people in the UK have to take their cases to county courts, employment tribunals, or to the Equality and Human Rights Commission (preceded by the Disability Rights Commission) to prove they were discriminated against? Morris (1991), for example, writes about disabled people who have experienced hardship and public struggles, and the toll this takes on the individuals and their families.[12] Finally, as for leaving rights to the realm of individual choice, this still positions disabled people as consumers.[13] Within the liberal-distributive framework, we are all consumers. As noted earlier, Stiker (2000) contends that all individuals in Western societies are measured in terms of their economic value as workers and consumers. Without consciously questioning liberal tenets of law, and creating

---

11   Indeed there is evidence to suggest that the Disability Living Allowance, for example, does grant disabled people more choice in accessing services in their day-to-day lives. See www.scope.org.uk.

12   In Chapter 2, I elaborate on how the process of taking up DDA cases can cause harm to disabled people.

13   See Chapter 3 for critiques of the DDA Part 3's consumer-oriented model.

new laws outside of this framework, disabled peoples' lives will not be truly transformed (Oliver and Barnes, 1998).

Critics have questioned the paradigm of the liberal citizen as an independent, able-bodied self-actualising individual (Oliver, 1990). Young (1990) shows that one's dependency does not necessarily have to imply that one is not a full citizen. She questions, as do other feminists, the assumption in liberal societies that in order to be full citizens, people must be independent and autonomous (ibid.). This view is seen to be individualistic and to have originated from a male experience of social relations, which values solitary achievement (Gilligan, 1982, in Young, 1990). In liberal societies, which value autonomy, justice is seen as giving people the opportunity to be independent. A different concept of justice, according to some feminists, would accord decision-making power to all, including those who are dependent and independent (Ramsay, 1997). As Young (1990:155) notes:

> Dependency should not be a reason to be deprived of choice and respect, and much of the oppression many marginals [sic] experience would be lessened if a less individualistic model of rights prevailed.

In light of these limitations, any law seeking to overturn discrimination against disabled people would have to challenge accepted norms and distributive ideas of justice (Young, 1990). How is law viewed as a social tool, and what expectations are placed on the law? In American society in particular, citizens place much faith in the power of law to redress social imbalances (Blomley et al., 2001). As I demonstrate in Chapter 2, this influences how the DDA compares to the ADA in its aims and ambitions. As Blomley et al. (2001:108) note, "compared to the English tradition, it is also apparent that many [American] citizens have remarkably high expectations of the law to penetrate through to the practice of everyday life." Gooding (1994:29), however, warns that one must not "ask too much of individual pieces of legislation, by abstracting them from the surrounding context of broader social and economic changes."

Critical legal scholars argue that this view of the power of the law to change society has led to a hegemony of law (Blomley et al., 2001). They note that hegemonic law is the decisive factor in social life and in determining human relations (ibid.). Critics argue that law supplants and suppresses local knowledge, individual experience, and notions of personal difference (Gilligan, 1982, in Young, 1990 and in Gooding, 1994). Foucault (1979) theorises that hegemonic law is empowered by the bureaucratic state's suppression of individualism. As Blomley et al. (2001:111) note:

> The ambition to structure social life and control the logic of judicial decision making with reference to an aesthetic theory of the law ... aims to replace the politics of local life with the logic of law.

Blomley et al. (2001:112) also argue that,

> to the extent that language is both representative and a constituent part of how
> people understand their lives, the hegemony of the language of law may deny
> the legitimacy of their personal experience.

Blomley (2000) notes that there is a growing body of scholars, as well as activists, who believe that the institution of law itself is at odds with the interests of citizens. Much has been written about the need to go beyond law in order to transform societies into being more equitable and just (Razack, 1998; Young, 1990). Foucault (1979:60), for example, argues that changing the system will require surpassing the legal remedies of state apparatuses:

> One of the first things that has to be understood is that power isn't localized in
> the State apparatus and that nothing in society will be changed if the mechanisms
> of power that function outside, below and alongside the State apparatuses, on a
> much more minute and everyday level, are not also changed.

Before concluding this section, I provide an example of how law, in liberal terms, has limited capabilities to overturn discrimination. ADL, for example, cannot reach the realm of interpersonal interactions and cultural attitudes if it is limited to the distributive paradigm of justice. The portrayal of disabled people in the media and in mass culture, for example, is not addressed in ADL (Morris, 1991). Young (1990:20) claims that discrimination occurs in the form of cultural oppression and notes that "such outrage at media stereotyping issues in claims about the injustice not of material distribution, but of cultural imagery and symbols." This view is shared by many, including hooks (1992), in *Black Looks: Race And Representation,* and Razack (1998), in *Looking White People in the Eye.*

hooks (1994), Razack (1998) and Young (1990) argue that we cannot legislate people to change their feelings towards disabled people and those seen as "others." Discrimination and prejudice become all the more challenging because of their more subtle and covert nature. For example, Alexis de Tocqueville (1835) noted in his seminal study on social culture in America that customs are more important than laws.[14] Customs are examples of powerful non-legal norms (Altman, 1990). There is a relationship between legal relations and social relations, but the precise nature of that relationship is contentious (Fitzpatrick and Hunt, 1987). Young (1990:130) says that following the struggles for equal and civil rights since the 1960s,

---

14   De Tocqueville's study, *Democracy in America* (1835 and 1840), was seminal because of the vast empirical scale of this work, written in two volumes, in which he travelled across the USA conducting in-depth research on American customs and attitudes. It was also the first significant study of the subject matter.

legal and social rules now express commitment to equality among groups, to the principle that all persons deserve equal respect and consideration, whatever their race, gender, religion, age, or ethnic identification.

Young (1990) argues that racism, sexism, ableism and homophobia exist in Western societies both explicitly and also in less obvious forms. In Chapter 6, for example, I explore these ideas in the context of disabled students' experiences of bullying, developing Young's (1990) argument that overt discrimination and violence towards disabled people have been sublimated into more opaque and covert forms of discrimination such as bullying, verbal abuse and exclusion. Young (1990) notes that while explicit discrimination and exclusion are forbidden by liberal societies,[15] inequalities are pervasive in more implicit forms. Young (1990:124) says,

> Our society enacts the oppression of cultural imperialism to a large degree through feelings and reactions, and in that respect oppression is beyond the reach of law and policy to remedy.

It can be argued that laws seeking to overturn discrimination against disabled people need to address not only discrimination but also domination and oppression, which are much pervasive and yet subtler than overt discrimination[16] (Matsuda et al., 1993; Razack, 1998; Young, 1990). By overt I mean classifiable, quantifiable, measurable and provable (Ward, 1998). Domination and oppression negatively impact the lives of oppressed groups such as disabled people (Young, 1990). This view is supported by Gooding's (1994) contentions about the failure of the formal equality model. Gooding (1994) notes that formal equality laws are based on a simple model of tort law, i.e. person A causes harm to person B and therefore person A owes person B something to compensate for the harm. According to Gooding (1994), Razack (1998) and Young (1990), discrimination does not work so simply but can be more subtle and yet less directly causal.

In this section I have examined the scope of law and its limitations in terms of its ability to change social inequalities. I have illustrated arguments that law in liberal societies is limited, including the contention that it is influenced by a *Distributive Paradigm of Justice* (Young, 1990). I have also shown that there are certain areas of human relations, including subtle interpersonal forms of prejudice,

---

15   Although even the DDA explicitly excludes some people from its protection if they are not being able to prove that they are 'disabled' over the long-term and that this 'disability' [sic] places them at a substantial disadvantage.

16   Domination is defined by the institutional conditions which inhibit people from participating equally in decisions and processes that determine and condition their actions. Oppression consists of systemic institutional processes that prevent people from "learning and using satisfying or expansive skills in socially recognized settings," or from communicating their feelings or perspectives on social life to others in society (Young, 1990:256).

which the law in its current manifestations cannot reach (Gooding, 1994, Matsuda et al., 1993; Morris, 1991, Razack, 1998; Young, 1990). In the next section I examine critical legal geographical theories and discuss their role in understanding disability discrimination law through geography.

(1.6) Space, Place and Specificity: A Critical Legal Geographical Approach

Within the critical legal studies framework, there are geographers who interrogate the legal from a critical geographical perspective (Blomley, 2000, 1994; Blomley et al., 2001; Clark, 2001; Pue, 1990). Blomley (2000) argues that through their critiques, unacknowledged assumptions about space – which work to stabilise the validity of legal propositions and identities and the meaning of law – are contested and revealed. Sarat (1997) posits that social reality is shaped and understood in terms of the legal, and furthermore, that it is also shaped and understood in terms of space and place (in Blomley, 2000). In this section of the chapter I locate this book within some of these geographical debates. In (a) I examine the importance of context, and in (b) I discuss geographies of disability.

(a) The Importance of Context

One of the key arguments of critical legal geographers is that Western legal systems have liberal foundations (Blomley et al., 2001; Blomley, 2000; Clark, 2001; Pue, 1990). One of the implications of legal liberalism is that those who write, interpret and enforce law have particular geographical understandings and assumptions about the world, despite liberal notions of abstraction and neutrality (Clark, 2001; Pue, 1990; Razack, 1991; Young, 1990). For example, theorists like Pue (1990) note that these geographical understandings cannot be epistemologically separated from the form and content of law. Pue (1990) argues that law in its Western modes is anti-geographical. He (1990:566) notes,

> Law is ... a profoundly anti-geographical faith. Judges are its high priests, courtrooms its sanctuaries, professional schools is seminaries. Its scriptures are "authorities" passed down from generation to generation by appointed oracles. Its god is a decontextualised, highly abstracted, and depersonalised "rationality." Contexts of all sorts – gender, class, religious, cultural, political, historical or spatial – are the enemies of Law. In all its majesty Law is the antithesis of region, location, place, community.

While Pue's (1990) views are particularly strong, it can be argued, as noted in (1.4), that Western law has traditionally been conceived of, and controlled by, able-bodied heterosexual white men, and has sometimes ignored other contexts and perspectives (Minow, 1991; Razack; 1998; Young, 1990). One of the aims of this book is to illustrate the importance of context in understanding disability discrimination law. The DDA operates in particular social, legal and political contexts which need to be examined, as they can influence the impacts of such laws on disabled people. I examine such contexts, for example, in Chapter 2, in which I outline the DDA in the UK and provide some comparison to equivalent laws in the USA and Australia. I also consider how the DDA is unfolding in the context of the higher education sector in Chapter 4. In Chapters 4, 5 and 6 I argue that HE in the UK can be viewed in the particular context of managerialism, defined as the application of ideas and principles from private-sector management, such as rationalisation and modernisation, into the public sector[17] (Newman, 2001).

As discussed in (1.5), these contexts – of liberal legal systems, the formal equality model and rights discourses – are seen by some theorists as being limited, narrow, and even disabling, and they question the efficacy of laws like the DDA (Barnes et al., 2002; Gooding, 2000; Oliver and Barnes, 1998). Rather than dismiss the ADL's efficacy altogether, the DDA and its chances for creating inclusion can be evaluated by placing the legislation in its particular contexts and in seeking to understand how law is mediated by social, cultural, and institutional factors. For example, in Chapter 5 I argue that the DDA is influencing changes in higher education, but that these changes are part of wider institutional cultures and are thus mediated by local actors, contingencies and specificities.

If critics like Pue (1990) argue that law is acontextual and anti-geographical, I demonstrate that the interpretation and implementation of law are profoundly geographical and linked to the local, the specific, and the particular. Blomley et al. (2001) argue that specificity, and the importance of place, ought to be central to legal principles. Blomley (2000) notes that law is geographical, envisaging the world in very particular ways, whilst possibly ignoring or masking these processes of envisaging. An aim of this book is to unveil these processes, and to inquire about the ways in which legal actors envisage the production of social and geographical spaces.

One way in which lawmakers, lawyers and judges envisage the world is through utopic and idealised liberal ideals of freedom, justice and equality (Blomley et al., 2001). In liberal legal systems, all citizens are thought to be free, equal and to be treated in the same ways by the state (Ramsay, 1997; Young, 1990). As illustrated in (1.4), such idealised and utopic views can ignore systemic inequalities and disadvantages (Gooding, 1994; Razack, 1998). As Blomley (1994:xi) notes:

> Legal interpretation, for example, with its encoded claims concerning the "utopic" location of the individual legal subject and the assured divide between universal and particularised legal knowledge implies, in this sense, a claim concerning the acontextuality of legal interpretation. The assertion of legal

---

17   Because institutions accredited by the UK government all receive public funding, they are nominally part of the public sector, although they operate as quasi-independent institutions. This is very different from the USA, where there are both public and private universities.

closure constitutes not only a rejection of the historicity of social life but also of its spatiality.

I employ critical legal geographical tools of re-inserting the importance of space and context into legal debates as a way of countering hegemonic ideals of law (Blomley, 1994). Pue (1990:572), for example, says that "any geographical approach to law is insurrectionist." While I am cautious in making such claims, I utilise critical legal tools such as deconstructing legal discourses and situating law in its social contexts. Blomley (2000:14–15) explains these tools when he says:

> The power of the critical onslaught..is that it is able to engage mainstream legal scholarship on two concurrent fronts. Given the institutional proximity of critical legal scholars to the bastions of the law, one line of attack is "internal," with an attempt at unravelling and contesting legal discourse on its own terms. Critics have borrowed from deconstruction, for example, to argue that the liberal worldview that underpins law is far from coherent, but is beset by internal contradictions and the associated repression of these contradictions..Another line of attack..is "external." Rather than focusing on the internal contradictions of legal discourse, the attempt here is to challenge legal closure by situating law: that is, by exploring and challenging its claims concerning law *in* society. In so doing, the very boundaries that separate law from society (and make possible my distinction between "internal" and "external" critique) are made unintelligible.

I also use critical legal geography to illustrate hegemonic power relations which favour able-bodied people (Gleeson, 1999). Foucault's (1979) theories about the complex relationships between power and space have been used by numerous geographers (Cloke et al., 1991; Radcliffe, 1999; Imrie, 1997; Blomley, 2000; Massey and Allen, 1999). For example, his theories on governmentality – defined as an attitude, ethos and mentality which pervades all actions[18] – can be used to illustrate the complex and varied ways in which institutions respond to the DDA, and the hierarchical power relations involved in institutional decision-making (Foucault, in Dreyfus and Rabinow, 1983). While institutional decisions and actions may be conducted in the name of efficiency or rationality, there are sometimes subtle power dynamics in which institutional actors create policies and processes regarding disability discrimination whilst subverting or obfuscating real attempts at inclusion by disability officers[19] (Barnett et al., 2002).

---

18   For more detail on Foucault's (1979) theories of governmentality, see Chapters 4 and 5.

19   For example, one disability officer noted, in survey, that her senior managers were excluding her from providing input into disability decision-making processes claiming that they were creating new procedures in response to the DDA. She then found that whilst claiming to implement the DDA, they were being less inclusive of disabled students. See Chapter 5 for more detail.

In this section I have discussed my use of critical legal geography to insert ideas of context and specificity into legal discourses. Using the particular example of higher education, I illustrated that context matters and that geography, in situating and locating specific institutional attitudes and practices, can help to reveal the diffuse power relations and governmentalities of higher education institutions (HEI) (Foucault, 1979). In the next section I discuss my engagement with geographical theories concerning the material experiences of disabled people.

(b) Geographies of Disability

As noted in (a), in examining how the DDA is being interpreted and implemented at the institutional level, it is important to contest the abstract neutrality of law because, as critical legal scholars posit, law is relational and acquires meaning through social action (Clark, 2001; Blomley et al., 2001; Pue, 1990). Describing how disabled people navigate their way through the attitudes, values and practices of higher education and service-providing institutions gives us insight into aspects of their social and educational lives which able-bodied people rarely obtain. The unfolding of disabled peoples' narratives can help contribute to our knowledge of the lived realities of disability and can be added to the plethora of experiences and contexts which feed into broader social and legal debates (Holloway, 2001; Konur, 2000).

I therefore use geography in this book because geography as a discipline seeks to understand the world, its people and places, and how they relate to each other (Blomley et al., 2001; Cloke et al., 1991). This relatedness is important in the legal realm, which Pue (1990) and Blomley (2000) criticise for ignoring contexts of people and place, as discussed in (a). For example, as Cloke et al. (1991:192) note:

> We are attracted to the suggestion that combining this distinctively postmodern sensitivity to different *people* with our long-standing geographical sensitivity to different *places* signals one (though not the only) road to travel in forging a valuable and still *critical* human geography ..

Such critical geographical theories can be useful in seeking to understand the experiences of disabled people. Gleeson (1999:195), for example, argues against reductionist or medicalised theories of disability in which disabled people are assumed to be "a simple reflection of natural or social forces." Gleeson (1999) calls for a geographically informed model of disability which includes the ways in which disabled forms of embodiment are socially produced and mediated, whilst acknowledging the material realities of impairment. These are some of the geographical ideas with which I engage in Chapter 6, for example. I demonstrate that disabled students' experiences are shaped by both their material experiences of impairment as well as broader socio-institutional forces which are specific to their particular geographies and locales.

Gleeson (1999), in his book *Geographies of Disability,* explores the ways in which social and physical spaces can work to disable people with impairments. These ideas are salient to this book for two main reasons. First, according to Gleeson (1999), notions of space, mobility and accessibility are significant in understanding disabled peoples' experiences. Second, disability has historically been given relatively little attention by "most western social scientists, including those in the spatial disciplines, Urban Planning, Geography and Architecture" (Gleeson, 1999:1). He notes that

> Disability is ... a vitally important human experience that Geography cannot afford to ignore. A failure to embrace disability as a core concern can only impoverish the discipline, both theoretically and empirically (ibid.).

In addition to Gleeson's book, there has been much more writing which addresses issues of disability and disabled people in relation to geography (Butler and Parr, 1999; Chouinard, 1999; Imrie, 2004a; Imrie and Hall, 2001; Kitchin, 2000). In considering relationships between space and disability, Gleeson (1999) and Imrie and Hall (2001) argue current accessibility regulations are inadequate at providing inclusive environments for disabled people[20] and that several socio-geographical factors, such as disablist attitudes and exclusionary socio-political structures, undermine policy environments. As Gleeson (1999:196) notes:

> The neo-liberal political-economic agenda, in particular, has grave consequences for marginalized social groups, including disabled people. Neo-liberalism has the demonstrated ability to erode and/or distort progressive social and environmental policies in ways that worsen the injustices experienced by disabled people.

In employing theories about geographies of disability, my aim is to produce an account of how certain institutions conceive of disability and the DDA, and how these conceptions impact disabled people. In so doing, I highlight Gleeson's (1999) theories on embodied historical-geographical materialism, which he shortens to 'embodied materialism.' As Gleeson (1999:32) argues,

> the best way of approaching disability, theoretically and empirically, is through a broad historical materialist framework which foregrounds embodiment as a key dynamic through which human societies are produced.

---

20   "An inclusive environment is one that can be used by everyone, regardless of age, gender, ethnicity or disability. It has many elements such as societal and individual attitudes, the design of products and communications and the design of the building itself. It recognises and allows for differences in the way that people use the built environment and gives all of us the chance to join in mainstream activities equally and independently, with choice and dignity" (DRC, 2005). For more on inclusive environments, see also Imrie and Hall (2001).

He feels that approaching an understanding of disability from a materialist account of the body has the advantage of constantly foregrounding the fact that impairment is only one of a range of overlapping embodiments, including those defined by sex, gender, race and class. I engage with Gleeson's (1999) theories cautiously as he seems to approach sex, gender, race and class as taken-for-granted categories without defining, qualifying or situating these terms in the same manner that he does the term 'disabled people.' I also want to maintain the focus on able-bodied norms and social practices, and therefore, while I employ Gleeson's notions of the materiality of disabled people's experiences, I do not 'constantly foreground' my discussions of disability in impairment per se (Barnes et al., 1999).

In ending this section of the chapter, I agree with Gleeson's (1999) argument that geographers have the potential to develop a fuller and more emancipatory notion of disability.[21] While Harvey (1996) notes that theoretical explorations into the production of space have been overlooked in some historical-materialist traditions, there is some evidence that non-geographers have critically engaged with issues of space (Durkheim, 1964; Lefebvre, 1991). There are also works such as *Race, Space and the Law: Unmapping a White Settler Society*, written by Razack (2002), an educational theorist, which engages with law's role in the production of racialised space and draws on theories from geography, law, and cultural studies. While I could have constructed this book exclusively around geographical theory, given its wider implications for law and policy, I have employed an interdisciplinary approach, drawing on the work of geographers, legal theorists, educational theorists, political scientists, feminists and philosophers (Barnett, 2002; Blomley et al., 2001; Foucault, 1979; hooks, 2003, 1994; Hunt and Wickham, 1994; Minow, 1991; Newman, 2001; Razack, 2002; Young, 1990).

In relation to my empirical study, I employ geography to demonstrate the importance, and gain an understanding of, spatial difference as it relates to disabled peoples' lives. Throughout the book, I emphasise the importance of context and specificity in opposition to liberal legal notions of neutrality and abstraction from difference. The empirical study which forms the basis of this book also uses geographical notions of *scale* to highlight the multi-faceted and complex relationships between different geographical contexts, in the same way that Jacobs (1996) highlights the trajectory of scalar relationships between empire, nation-state, city and locale in *Edge of Empire*. Specifically, the book examines disability discrimination law in various scales: across international comparative contexts, at the level of the nation-state, within regions, at the institutional level, and in the life contexts of individual disabled people. In the next section I conclude the chapter.

---

21   See Clare (1999), *Exile and Pride* for a discussion of emancipatory notions of disability.

(1.7) Conclusions

In considering the efficacy of disability discrimination law in overturning discrimination against disabled people, many challenges arise. One of the main ideas this book considers is in what ways legislation can change social actions and behaviours to be more inclusive. Legislation as a tool can help to remove barriers, can empower people to legally seek their rights, and can even punish those who discriminate, but ultimately legislation is but one tool in a complex array of individual behaviours, social norms and cultural expectations. In addition to law,

> what is needed is an alternative language of social life and an alternative means of understanding our obligations to one another outside of the language of law. The institution of law seems to have taken over the realm of democratic practice (Blomley et al., 2001:114).

Viewed more critically, law can be seen as an active constitutive force through which one culture establishes its identity by rendering another culture unfit. In this cultural struggle then, disabled people have to assert their value and worth, not just by using legal tools, but also by gaining social acceptance and inclusion. As Kobayashi (1990) notes, one of her objectives in looking at the social constructions of law is

> to recognize the law beyond its institutional process, that is, to go beyond the exercise of connecting the law and society, to recognizing the law *in* society, as part of the structured system it helps to engender. Legal and social relations are thus mutually penetrated, in a discourse (in Pue, 1990:575).

It is critical that disabled people, in attempting to use disability discrimination law as an empowerment tool, recognise that these laws are intertwined with the legal and political cultures of their nation-states. Thus it would not be prudent to expect that all of society's prejudices and conceptions about disability can be overcome due to an act of law.

Oliver and Barnes (1998) feel that the central value system and ideological foundations on which western society rests remain unchallenged by ADL and that only such challenges will lead to the inclusion of all people – regardless of race, gender or ability. Whether or not such a fundamental challenge is achievable or feasible, they see a need to go beyond the remit of ADL in order to achieve an inclusive society:

> The route to inclusion is through complete social transformation and not simply in terms of the extension of the idea of civil rights to disabled people (Oliver and Barnes, 1998:56).

It is not my aim to draw simplistic conclusions that the DDA is comprehensive and complete, in terms of its ability to redress discrimination against disabled people, or useless and incapable of effecting such change. To do so would be reductionist. Instead I have examined these laws, with (1) a critical legal geographical approach, and (2) an understanding of how law is conceived in liberal societies. As such, I have illustrated the ideological frameworks behind these laws, i.e. conceptions of the legal subject generally, and of disability specifically. I have examined arguments about law's efficacy and raised questions about how law can transform society's disabling foundations. The rest of the book further develops arguments about law and legal liberalism and explores disabled peoples' experiences of these laws and legal processes.

Chapter 2

Law, Geography and Disability: Revealing the Idealised Spatialities of Disability Discrimination Legislation

(2.0) Introduction

As outlined in Chapter 1, this book examines disability discrimination law with the goal of understanding its impact on disabled peoples' lives. While its primary focus is on the UK, this chapter makes comparisons with, and connections to, similar legislation in the USA and Australia.[1] All three countries have liberal legal roots, and there has been much sharing of legal information and language, and cross-pollination of ideas and concepts relating to disability discrimination. By analysing the language, structure and impact of the DDA – with reference to the DDA 1992 in Australia and the ADA 1990 in the USA – this chapter highlights the variety of outcomes for disabled people and makes comparative analyses of the effects of these laws.

This chapter employs empirical material from interviews with key players in the field of disability discrimination legislation in the UK, USA and Australia, and analyses the legislation and some of the case law to date. The chapter is organised into the following sections: (2.1) Liberal roots of disability discrimination law, (2.2) The role of legal context in framing disability discrimination legislation, (2.3) The DDA: language and assumptions, (2.4) Legal abstraction and the importance of context, and (2.5) How law harms: disabling aspects of the DDA. In (2.6) I conclude the chapter.

(2.1) Liberal Roots of Disability Discrimination Law

Drawing on the work of critical legal theorists in geography (Blomley, 2001; Clark, 2001), and education (Razack, 1998; Williams, 1997b), this chapter examines the notion of rights, resting as it does on liberal foundational notions of equality. Expanding on the discussion of rights in Chapter 1, this chapter disentangles these

---

1   In the context of this thesis, Chapter 2 analyses some of the experiences of similar legislation in Australia and the USA; Chapter 3 provides a more detailed analysis of the DDA (UK) in relation to service provision, and Chapters 4, 5 and 6 provide in-depth analyses of the higher education provisions of the DDA (UK).

laws from their particular social, political and historical contexts and to unearth the assumptions, values and contingencies embedded within them. This is what Blomley et al. (2001) call the project of destabilising the law by revealing its propositions, identities and meanings. As they note,

> Unacknowledged assumptions about space that work to stabilize the validity of seemingly obvious propositions, identities, and the very meaning of "law" are revealed. And, to the extent that providing this stability is a function of their *remaining* unacknowledged and unexamined, then this apparent stability may be called into question or revealed as contingent (Blomley et al., 2001: xv).

As suggested in Chapter 1, linking geography to law, Blomley et al. (2001) argue that specificity, and the importance of place, are central to legal principles. They note that law is geographical: it involves boundaries, territories and the governance of various aspects of people's everyday existences, all of which occur in particular places (ibid.). In the context of disability discrimination legislation, in what ways these laws envisage space, and what geographies they reference, may impact their outcomes for disabled people. What are the imagined spatialities of these laws? For example, does the DDA adhere to what Blomley (1994) calls utopic liberal notions that we are all free and equal, or does it acknowledge the systemic spatial disadvantages experienced by disabled people?

Antidiscrimination discrimination law (ADL) and civil rights law (CRL) draw on two philosophies, both of which impact their effectiveness. First, this legislation draws on utopic liberal ideals of equality and justice, premised on the belief that rational autonomous citizens adhere to law,[2] and that discrimination can be adequately addressed ended by legislation.[3] As Razack (1998) reveals in *Looking White People in the Eye: Gender, Race and Culture in Courtrooms and Classrooms*, this is a fallacy. She notes that law relies on a positivist conception of knowledge, which assumes that discrimination is overt, quantifiable and measurable (ibid.). Creating legal equality without adequate attention peoples' daily, lived experiences – including those of disabled people – does not fully address these experiences, and allows law to operate at an idealistic level, some distance from social, material and geographical contingencies.[4] As she (1998:23) notes,

> The daily realities of oppressed groups can only be acknowledged at the cost of the dominant group's belief in its own natural entitlement. If oppression exists,

---

2   Or other laws which attempt to modify behaviour or attitudes.

3   The goals of the ADA and DDAs are to end discrimination but not oppression. The way this frames the 'problems' faced by disabled people already limits the legal remedies for them.

4   As Justin Dart, one of the authors of the ADA, said, "rights don't buy you bread at the supermarket." This is an example of theoretical rights versus the practical realities of life. (Quoted by an interviewee in America, March 2003).

then there must be oppressors, and oppressors do not have a moral basis for their rights claims. If, however, we are all equally human, with some of us simply not as advanced or developed as others, then no one need take responsibility for inequality.[5]

Using Razack's (1998) arguments, to be effective, ADL and CRL would need to acknowledge the role of able-bodied people in creating able-bodied privilege, and maintaining social systems which oppress disabled people. To what extent laws succeed, or not, at focusing on the role of able-bodied people in maintaining systems which oppress disabled people is an idea explored in this chapter.

In thinking about how accurately law and legal ideals reflect the lived realities of disabled people, a second argument illustrates that law can and does impact on disabled people. The category of 'disabled' is itself is a legal construct which has material implications for disabled people, and, as noted in Chapter 1, combines two elements of experience – the social and the embodied (Gleeson, 1999). So while the first argument about law is that it relies on utopic ideals of equality which do not necessarily reflect reality, the second is that this legislation can itself be responsible for creating material inequality and oppression through its use of categorisation, labelling and segregation, or what Said calls processes of 'othering' in his seminal work, *Orientalism* (1978). Said (1978) and Foucault (1991) have both argued that law can be seen as constitutive of social- and therefore self-consciousness, e.g. legal categories such as 'disabled' cause disabled people to see themselves as 'other.'[6]

Throughout the chapter I use the empirical evidence to reveal law's contingencies and assumptions about disabled people. Fitzpatrick & Hunt (1987) and Roulstone (2003) posit that law has in-built assumptions and prejudices which favour able-bodied people, and which are so deeply embedded in juridical-legal mindsets that they are not immediately apparent and need to be unearthed. These mindsets will be interrogated throughout the chapter, and in particular, the examination of case law will reveal how these able-bodied assumptions work in the context of courtrooms. As Blomley et al. (2001: xvii-xviii) note,

> Taking legal themes such as the connection between law as discourse or representational systems and law as power seriously opens up a variety of questions about how – by what actual practices, in relation to what social or political projects – social space is produced, maintained or transformed.

As Razack (1998) notes, this process of unearthing legal practices, projects and assumptions reveals law's liberal roots in the Western world. She notes that

---

5   For further discussion of the differences between discrimination and oppression, see Chapter 1.

6   For more detailed analysis of the categorisation of disabled people in law, see Chapter 6. Foucault's reference can be found in Razack, 1998:33.

contract theory is central to all legal principles[7] (ibid.). Contract theory is based on the idea that we are all bound to the state and each other through a series of contractual obligations (Williams, 1997b). When discrimination against disabled people is acknowledged, one approach to ending this discrimination is to create contracts in which disabled people ought to be given access to the same privileges as able-bodied people. But because contract theory focuses on individuals and their relationships to the state and does not look at collectivities or groups, this model fails to adequately tackle discrimination (Razack, 1998).

For example, disabled pupils in England are meant to be 'statemented' so that their needs are met by local educational authorities (LEA); in practice, while this 'contract' meets the pupils' needs, it does not remove the bullying or social stigma which can accompany the formal identification of impairment in a child. Given the origins of the liberal ideals of equality and justice, rooted as they are in idealistic principles, Razack (1998:30) asks, "Can we use these constructs without immediately relying on their built-in relations of domination?"

These liberal ideals are based on definitions of human beings as rational legal subjects pursuing their individual self-interests. As noted in Chapter 1, this is perhaps not the most effective formula for facilitating the inclusion of disabled people (Ramsay, 1997; Young, 1990). Williams (1997b) argues that while Western legal subjects are entitled to various social contracts, we are also bound by liberal forces which are relational and contingent.[8] Simply acquiring legal civil rights does not eliminate these complex forces and power relations and nor does it guarantee the fulfilment of such contracts. As Foucault (1980) notes, power is not something one "possesses, acquires, cedes through force or contract" but power is a relation of force (in Razack, 1998:33–4). Therefore codifying civil rights, without paying adequate attention to broader issues of power can mask the domination and oppression of disabled people whilst engaging in a legal rhetoric of freedom and equality. As Razack (1998:35) notes, "we need to ground the discussion of collective rights in concrete social realities." How far these three laws are related to, grounded in, and engage with social realities will be key to understanding their efficacy in ending discrimination against disabled people, which is an ongoing goal of this chapter.

---

7   The idea of contract theory is also related to Young's *Distributive Paradigm of Justice* as discussed in Chapter 1.

8   Burton (2011) implies that these social contracts are not adequately enforceable and argues that the UK should enshrine these social contracts in the form of legislation guaranteeing economic and social rights. See Burton, J. (2011) "It's time to enshrine socioeconomic rights in law", *The Guardian*, 28 October 2011.

(2.2) The Role of Legal Context in Framing Disability Discrimination Legislation

In January 2011, the UK government proposed to replace the Disability Living Allowance (DLA) with a Personal Independence Payment. This proposed change is for the purpose of increasing the efficiency of the current delivery of benefits to disabled people, and for the purpose of cutting 20 per cent of costs (BBC, 8 January 2011). Another change will be the requirement that claimants undergo new assessment tests and be required to have an impairment for a minimum of six months (ibid.). The Disability Alliance, a charity, says that these measures will lead to the removal of 380,000 individuals from the DLA payments. Another charity, Scope, said it disagreed with the proposed discontinuation of the mobility portion the DLA for those who reside in care homes. Scope's chief executive noted that it "is quite a callous decision. It will result in people being prisoners in their own homes, they won't be able to do those daily things that everybody else would take for granted" (Hawkes, R., 2011, quoted in BBC, 8 January 2011).

The current political context regarding the provision of disability benefits illustrates the material challenges disabled people face despite the existence of ADL. Going farther back, in 2005 the Department for Work and Pensions published its five-year strategy entitled: "Opportunity and security throughout life" which included a 'claw back' on the *Incapacity benefit* by increasing means testing and enforcing more stringent conditions for these benefits from 2008 in the hopes of coaxing more than a million claimants back into jobs (Wintour, 2005). Statistics already show that disabled people are much more economically marginalised than the average Briton, so the effects of these claw backs could be negative (DRC, 2005). As I argue in Chapters 4, 5 and 6, this is part of the government's managerialist drive to modernise Britain and has serious implications for disabled people. This section of the chapter, however, goes back to the pre-DDA years to help illustrate the contexts from which this law, and its American and Australian counterparts, were developed.

The contexts within which the ADA, DDA (Australia) and DDA (UK) were developed share similarities and differences. Broadly speaking, all three come from a liberal legal tradition and share Anglo-American roots, i.e. both the USA and Australia are former British colonies. However, the subtle differences between these countries' legal philosophies, governance systems and cultural assumptions can highlight different attitudes towards civil rights in general, and disability rights specifically (Campbell and Oliver, 1999).

The USA was the first country in the world to enact comprehensive legislation aimed at ending discriminatory practices towards disabled people (Gooding, 1994). In fact, even before the ADA (1990), which came into force in 1992, US Congress had enacted the Rehabilitation Act (RA) 1973. Section 504 of the RA made it illegal for the federal government, or any of its agencies or recipients of federal funding, to discriminate against disabled people.[9] Apart from the legal content of the RA, its very existence in America so long before anywhere else in the world speaks volumes about the national culture of the USA. The language of empowerment for disabled people, and the language of formal equality, have been ensconced in the American psyche for far longer than in either Britain or Australia (Gooding, 1994).

As is detailed later, there is a disjuncture between the language of formal equality and material realities. Gooding (1994), hooks (2002) and Young (1990) argue that there is a gulf between the rhetoric of rights and the realities of American life, and that despite legislation barring discrimination based on age, ability, race, gender or social class, inequalities pervade the American landscape. The human tragedy which followed Hurricane Katrina in 2005, for example, highlighted the ways in which the US government's official responses to the human catastrophe were purportedly delayed in part due to the race and social class of New Orleans' inner city residents (Younge, 2005a). This example shows that while the US government has been a leader in establishing rights legislation, it still struggles to match its citizens' lived realities to its legal aspirations.

Within the UK, struggles for the inclusion of disabled people, through legislation or other means, have been long and hard (Campbell and Oliver, 1999). At a time when the US government was forbidding discriminatory practices in the employment of disabled Americans, the UK's disability movement was still in the nascent stages of its politicisation.[10] Other struggles also occupied the national political landscape. For example, the arrival of waves of new immigrants in the late 1960s and early 1970s, during the post-independence periods of former British colonies, caused new strains and racial tensions in an increasingly diverse society. While such struggles still continue to the present day,[11] despite the existence of the Race Relations Act, for example, that period was particularly turbulent[12] (Adams, 2002).

America's culture of rights can be linked to its strong notions of individualism, dating back to the *United States Constitution* (1787) and the first ten amendments to the constitution, collectively known as the *Bill of Rights* (1791). Americans have worked hard to enshrine their rights over the past two and a half centuries. For example, some writers note that the right to practice religion in whatever ways

---

9   While Section 504 has been criticised for being 'weak', it was still the first law of its kind on a national scale. For example, this law applied to most higher education institutions as more than 90 per cent of HEIs in the USA receive federal government funding, including private institutions (*New York Times*, 2005).

10   For example, RADAR, a disabled people's pressure group, campaigned for DDA-type legislation from the early 1990s onward (RADAR, 2003).

11   For example, Conservative Party Leader Michael Howard's immigration policy in the 2005 UK general election was labelled 'racist' by some (see Younge, 2005c).

12   An example of that turbulence and virulent racism can be found Enoch Powell's 'Rivers of Blood' speech to the Conservative Association in Birmingham on 20 April 1968, reproduced in the *Daily Telegraph* on 6 November 2007.

they wished was one of the primary reasons the first settlers, the Puritans, emigrated from England to the British colonies in America (Amar, 2000). Additionally, rights of representation and self-determination are cited as important reasons for the fighting of the Revolutionary War against Britain, and the rights of all people to be free – including slaves – was a key factor in the US Civil War being waged (ibid.). This long history of rights has influenced the disability rights movement in America. As one policy officer in Washington DC noted, in interview,

> America is a melting pot; it is more accepting of new ideas. It is easier to travel in the US and find environmental and attitudinal acceptance – more than anywhere else in the world (Director of a disability technical assistance centre).

At the same time, this inclusion through nationalistic notions of equality has its limits and contradictions. As one disability rights expert in Boston said, in interview,

> This is America; we treat all Americans well. Or at least, that is what the average American thinks. But the reality here in the US is that it is necessary to confront a prevalent perception that laws like the ADA have resolved the issues (Manager of a regional ADA assistance centre).

Likewise, an interviewee in Sydney noted that disabled people benefit from, and are excluded by, Australia's broader socio-political culture. He said:

> In terms of progress since the DDA, Australia has messy contradictions. The government has hegemonic ideas about social policy, and a very limited view of people with disabilities. Australia is some considerable way behind in the politics in some areas – for example, it has dreadfully bad gender politics, it is the most macho place on earth – but there is a willingness to embrace cultures of disability, Independent Living Movements, etc. (Head of a physical disabilities organisation).

In addition to the intertwining of race, gender and disability politics identified by the interviewee, changing gender roles also caused conflicts in UK political spheres, as did economic turmoil such the oil embargo, the winter of discontent, old infrastructure and increasing national debt (Dearlove and Saunders, 2000). While the government did enact the Race Relations Act and the Equal Opportunities Act, it would take another 20 years until the introduction of an antidiscrimination law for disabled people, the DDA.[13] As one British interviewee noted:

> Disabled people are 20 years behind women and racial minorities in the UK. For example, when the Commission for Racial Equalities released the Thurston

---

13   There was the Disabled Persons Act (1944) but that was based on paternalistic welfare model and did not accord 'rights' per se.

report, following the inquiry into the death of Stephen Lawrence, one of its recommendations was a duty on public authorities to promote racial harmony. They amended the Race Relations Act – so that such authorities would have to actively ensure they are doing this. No comparable requirement exists for disabled people. The law does not approach disability issues from a positive angle, but from a negative one (Policy officer at a government commission).

This is confirmed by Roulstone (2003:118) who notes that "disabled people and their allies had waited 20 years for the alignment of disability discrimination law with the much earlier Sex Discrimination Act (1975) and the Race Relations Acts of 1965, 1968 and 1976." The UK's political landscape was strongly influenced by its Prime Minister from 1979–1990, Margaret Thatcher, who sought to reduce the role of government in public life and to allow people to carry on with private enterprise with minimal state interference (Rose, 1987). This philosophy led to massive privatisation of state assets such as housing, services such as railways and utilities, and public institutions. The impact of these changes on disabled people was considerable, from formerly affordable subsidised housing being sold, to the de-institutionalisation of disabled people and policies of care in the community (Oliver & Barnes, 1998). One interviewee claimed that the enactment of the DDA by a Conservative government impacted its effectiveness and efficacy. He said:

There were thirteen attempts at a Disability Discrimination Act before 1994, which were mainly private members bills in parliament. Seven of the thirteen were by a Labour MP Tom Clarke. The twelfth time, the conservative government 'talked out' the bill to kill it off. The outcry which followed this, and the subsequent embarrassment of the government, led to the DDA initiative. However as it was written by Tories, it aimed to minimise the burdens on private industry, the Conservatives' biggest supporters (Policy Officer for a large public transport organisation).

Thatcher and her successor John Major encouraged policies which strongly reflected the capitalist free-market attitude of Ronald Reagan, US President from 1980 to 1988.[14] Both Thatcher and Reagan seemed to have little faith in public services and conceptions of the 'public good,' and both chose to reduce funding for government institutions[15] and services to the point where, for example, some Americans with psychological impairment were turned out onto the streets

---

14   Due to pressure from the powerful disability lobby, which included disabled Vietnam veterans who had become national heroes, discussion of an ADA started in the late 1980s, after Reagan was replaced by George H. W. Bush, a republican but slightly more progressive President than Reagan. As noted in Figure 4.1, the ADA was introduced in 1990.

15   Although, conversely, in the eight years during which he served as President, the US national debt grew from $1 trillion to approximately $3.5 trillion.

(Gewirtz, 1997). In seeking to reduce the role of government, Thatcher is quoted as saying "I believe that the great mistake of the last few years has been for the government to provide or to legislate for almost everything" (in Wapshott and Brock, 1983:279). This kind of thinking has influenced DDA rights. For example, disabled people have to meet obligations, and prove they are disabled in order to have entitlements (Roulstone, 2003).

Given all the social and political turbulence of the 1970s and 1980s, how was America so far ahead of Britain by introducing the RA in 1973? Some of the impetus for this legislation came out of post-World War I optimism and a desire to re-build after the great depression. US President Franklin Delano Roosevelt (1933-1945) envisioned the *New Deal Era* and employed Democrats such as Clark Foreman to implement his vision, which bolstered the welfare state and concomitant role of the government in helping people (Sullivan, 1996). Roosevelt was disabled, and used a wheelchair in all the years of his presidency (Struck, 1997). Subsequent governments built on the *New Deal Era* with policies creating more equality for men and women, and funding education and Medicare, for example (Sullivan, 1996).[16]

The socio-political context in the UK in the 1970s was quite different from the USA. For example, disabled peoples' rights did not feature prominently on the UK government's agenda at that time because of the particular ways in which 'provision' for disabled people had historically evolved in that national context. For example, Borsay (2004) details the elaborate programmes for the treatment of disabled people beginning during the Industrial Revolution and epitomised by the Chronically Sick and Disabled Person's Act (1970). Following the end of World War Two in 1945, and especially since the establishment of the National Health Service in 1948, disabled people in the UK were subjected to a medical-welfare bureaucracy state apparatus which included segregated schooling, institutionalisation and training workshops which sought to extract economic value from them (ibid.).

In the 1960s, a time when one could still find signs saying 'no blacks, dogs or Irish' in England, America produced Martin Luther King, Malcolm X and Gloria Steinem (Lydon, 2005). These differences can be attributed to many factors: the sheer size of the US, with a much larger population, including a substantial African American population, who were descendants of former slaves (Sarat, 1997). The UK had no comparable population of people of colour in terms of size and political clout and thus no natural hotbed of unrest and protest (ibid.).[17] As one Australian interviewee noted, issues of gender, race and disability discrimination are linked.

---

16   The RA 1973 was also inspired by the return of disabled Vietnam war veterans to the USA and the blossoming Independent Living Movement, founded in Berkeley, California in the USA.

17   Paul Gilroy (1991), in *There ain't no black in the union jack*, demonstrates the complexity of racial politics in the UK. He explores the relationships among race, class, and nation as they have evolved over the past few decades and challenges current sociological

He acknowledged the intersectionality of discrimination on various levels, and how these issues can affect individuals in multiple and complex ways. He said:

> Disability culture doesn't exist as a minority culture like race and gender. People with disabilities would like to be on par with other marginalized groups, although some Muslim or Chinese groups probably feel as excluded as disabled people ... I am aware of the fact that as a senior advocate in the disability community, I'm a 'white god with a beard'. There are only 2 women on the management committee of this organisation, out of 21 people. The previous president was also a 'white god with a beard'. There is the MDAA, the Multicultural Disability Advocacy Association. The disability community itself has some work to do. There was a view that the problems of people with disabilities are undifferentiated along lines of class, gender and race: "If you can't get into a building, you can't get in." There was a lack of subtlety – or acknowledgement of the complexity of power structures and discrimination. For example, 20% of the population we're supposed to represent is non-English speaking (Disability rights advocate).

Many of the American government's civil rights philosophies centred on guilt at the fact that more than a century after slavery had ended, the number of African Americans living in substandard conditions was disproportionately high (Gilroy, 1991). While immigrants from former British colonies did reside in the UK, there was not a large population of Britons who had descended from slaves and were living at substandard levels, as there was in American for many generations (ibid.). This stark difference between the quality of life based on race spearheaded the desegregation movement in the US, starting with the landmark Supreme Court decision in 1954 of Brown v. Board of Education, which outlawed the segregation of children in schools based on race,[18] and eventually paved the way for the Civil Rights Act of 1967 (Sarat, 1997).

While the US and UK governments of the 1980s were philosophically similar, as stated earlier, by the time of the late 1980s, the US Congress was considering new legislation with substantial power and the force to end discrimination against disabled people. The ADA (1990) took effect in 1992, the year Bill Clinton was elected US President, and his administration ensured follow-up, funding and enforcement of the legislation. Although the Act was signed by President George H. W. Bush, a Republican president, it represented the behind-the-scene efforts of the disability lobby in Washington D.C. (see footnote 14 for more detail). As mentioned earlier, there are strong links between US civil rights legislation and American historic notions of inalienable rights for all of its citizens. As Bush said,

---

approaches to racism which attempt to separate it from discrimination based on class, gender, sexual orientation, age and ability.

18   Like 'disabled', 'race' is a contested category and although presented here to highlight the basis for the decision in Brown v. Board of Education, and should not be taken as a prima facie acceptance of this term by the author.

in his remarks on the Signing of the Americans with Disabilities Act on July 26, 1990,

> Across the breadth of this nation are 43 million Americans with disabilities ... Three weeks ago we celebrated our nation's Independence Day. Today we're here to rejoice in and celebrate another 'independence day,' one that is long overdue. With today's signing of the landmark Americans for Disabilities Act, every man, woman, and child with a disability can now pass through once-closed doors into a bright new era of equality, independence, and freedom ... This historic act is the world's first comprehensive declaration of equality for people with disabilities – the first. Its passage has made the United States the international leader on this human rights issue ... Our success with this act proves that we are keeping faith with the spirit of our courageous forefathers who wrote in the Declaration of Independence: "We hold these truths to be self-evident, that all men are created equal, that they are endowed by their Creator with certain unalienable rights" (US DOJ, 1992).

The rhetorical flourishes Bush uses indicate the US government's massive psychological investment of hope in this legislation and symbolically link it with the country's foundational Declaration of Independence of 1776. Like the DDA in the UK and Australia, the ADA seeks to end discrimination in the context of individual cases (Gooding, 1994). Unlike the DDAs in the other two countries, the US Government has actively pursued cases of its own. For example, the US Attorney General has the power to do this, where s/he feels an individual has been discriminated against and the case is of significant value beyond its geographical area.[19] The US Department of Justice (DOJ) has also funded Legal Aid programs so that disabled people can go to court, has funded federally-sponsored institutions such as universities and colleges, and has reserved power, at the highest level, to prosecute if necessary (Gooding, 1994). While the UK introduced the DDA in 1995, the phased-in approach of this legislation meant that some parts of the legislation had no legal force until nine years later. For example, since 2004, all goods and services providers in the UK have had to make reasonable adjustments to their business practices and premises in order to accommodate disabled customers (DRC, 2001).

Disabled people in Australia also had high expectations when their DDA (1992) was introduced, but thirteen years later, many are disappointed (Handley, 2001). For example, the commonwealth government of Australia established the Human Rights and Equal Opportunities Commission (HREOC), a quango enforcement agency which does not have the clout of government power like the DOJ in America, and has been severely undercut by inadequate funding. The HREOC receives complaints about discrimination against disabled people from all states

---

19   The US is such a vast country that the DOJ attempts to show the importance of key local cases on a national scale.

and territories in Australia, and processes these complaints through mediation processes. While the DDA has had symbolic significance for disabled people, its impact is questionable since the actual number of complaints of discrimination received by HREOC has gradually declined (Handley, 2001). While the declining number of cases might be seen as a success, in that it could suggest that less discrimination is being occurring, government funding cutbacks have led to slower and more arduous hearing processes, which are believed to have discouraged disabled people from bringing their cases forward (ibid.).

Disabled people in all three countries have faced similar challenges in ge.tting governments to take seriously the need to adequately fund the enforcement of their disability discrimination laws. These challenges are connected through similar liberal conceptions of law in these countries, which are all experiencing a climate of 'value-for-money' (Zifcak, 1994). Having examined the socio-economic and political contexts of the three countries, as well as the social changes which spearheaded ADL and CRL, the next section looks at the form and content of the three laws.

(2.3) The DDA (UK): Language and Assumptions

While more detailed analyses of the legislation follows in subsequent chapters, I here provide a brief outline of the DDA (UK). The DDA aims to end disability discrimination and gives disabled people rights in employment, access to goods, facilities and services, education and transport. The Act defines a disabled person as someone with "a physical or mental impairment which has a substantial and long-term adverse effect on his ability to carry out normal day-to-day activities" (HMSO, 1995:1). Employers, service providers and educational institutions are found to be discriminating against disabled people if they treat them less favourably (without justification) than others because of their 'disability', or by not making reasonable adjustments (without justification) (ibid.).

Examining the above language and comparing it to its American and Australian counterparts provides a context for how disabled people are referenced and reveals some of the assumptions about disability behind these laws. All these laws rely on ideas of the "other" and categorisation. They all borrow from medical models of disability in that the solutions to disability discrimination are seen in terms of accommodating individual disabled people rather than conducting systematic reviews of ableist practices and systems of privilege.[20] In that sense, from their very definitions all three laws delineate disabled people as 'others.' These definitions, as limited as they are, are currently under attack in all three countries given that disabled people are often in the position of having to prove their disabled status

---

20   The US government did conduct some broader reviews but those are being reduced due to funding cuts.

within legal processes.[21] As Roulstone (2003) notes, this alone is disabling, dehumanising and can cause real harm to disabled complainants.[22]

In America and Australia, disabled people are referred to as people with disabilities (HREOC, 2003; US DOJ, 1992). This particular terminology was used because those who drafted the legislation did not want 'disabled' to come before the person and thus stigmatise them (ibid.). Thus a person with a disability is simply someone who has an impairment, but it is not his or her defining characteristic. As one American interviewee noted:

> We in the U.S. prefer to refer not to label people first, so the preferred terms here would be people with disabilities, students with disabilities, etc., rather than disabled persons. I realise that this can be perceived as an issue of political correctness, but I think in this case it is a point well made (Coordinator of disability services for a university in Connecticut).

The term 'disabled people' is used in the UK legislation, as opposed to the American term 'people with disabilities.' It seems that in the UK, lawmakers have acknowledged the role society plays in disabling people with impairments. In some ways this is a unique achievement, in the sense that such language is not used in either the ADA or the Australian DDA. While UK disability theorists prefer use of the term 'disabled people,' and this is indeed the term employed in the UK DDA, the legal definition of disability is still premised on the individual-medical model of disability (Oliver and Barnes, 1998). Indeed, in Section (2.5) I show how these remnants of the medical model play out in legal processes.

Beyond the debate about the terms 'disabled people' versus 'people with disabilities', the legislation in all three countries uses the language of 'reasonableness,' which has invariably been a source of legal complexity and ambiguity in determining what is considered 'reasonable.' For example, one American interviewee noted:

> People don't understand what 'reasonable' means. They say, "Just tell me what to do." We need to break the orientation towards rules, and open them to thinking differently (Manager of a disability technical assistance centre).

A British interviewee noted:

> It is not clear to anyone, especially not judges, what 'reasonable' means. There have been some very negative judgements – courts seem to err on the side of the goods and services provider (Policy Officer for a government commission).

---

21 The Australian DDA makes some mention of social exclusion and stigma but its legal mechanisms are primarily based on individual cases.

22 The issue of harm to disabled people caused by the DDA is further examined in Section (3.5).

And one British interviewee offered his view on how notions of 'reasonableness' could actually work against disabled people by lowering minimum standards,

> I mean I'm being cynical about it because you saw the stuff that I wrote on Australia, that's what happened. That's the reality, that's what happened out there. Everyone thought that legislation was going to solve all the problems, it was the last bastion of you know breaking down discrimination, all institutions got very twitchy about it. The first case went to court, the institution won. They thought, hang on a minute, you know all these people scaremongers telling us we're going to get dumped but actually the case law is now saying that this is 'reasonable.' In actual fact, it could work against disabled people because what it might do is set minimum parameters (Disability Team Leader for a higher education funding council).

At the heart of this debate is the challenge faced by each of these three antidiscrimination statutes: how to create equality for all under the law, whilst taking into consideration the particular contingencies and geographies of each disabled person's context. In other words, how can the law treat all people equally, whilst acknowledging that different people will require different legal remedies to counter this inequality? As such, the debates in all three countries have centred on notions of formal equality versus substantive equality, equality of opportunity versus equality of outcome, reasonableness, fairness, and ultimately, what constitutes a civil right. As an interviewee noted:

> There is a difference between civil rights legislation which they have in America, versus antidiscrimination legislation, where the onus is on the individual and the discrimination has to be more explicit. The type of legislation makes a difference. Civil rights legislation tries to change society, but antidiscrimination legislation only deals with individual cases (Policy Officer for a voluntary sector organisation).

Another interviewee concurred that the DDA (UK) does not confer rights on disabled people per se:

> With regards to transport, all new trains and buses have access requirements after certain cut off dates, but there is no right to be served, so this is meaningless. A driver can refuse to allow a disabled person to board even if it's an accessible bus (Policy Officer for a government commission).

The different terminology and semantic strategies used in these countries reflect the varying notions of identity and political correctness in differing cultures, as noted in the example about the use of the terms 'disabled people' and 'people with disabilities.' Critics have noted that the ADA is more powerful in that more institutions and individuals can be found to be liable and responsible for providing accommodations, and unlike in the UK or Australia, it is much easier to file suit in America (Gooding, 1994). As one Director of Disability Services for a major university in the USA notes,

> Do faculty members contest the academic accommodations I suggest? Yes, all the time. I sometimes say: would you like to go to court? They usually comply because of fear of being sued (Director of disability services for a university in New York).

This is another example of how a social, cultural or political difference in the legal system of a country affects the impact and scope of law. Likewise, the particular social and geographical contexts of Australia have informed how that country's legislation has been interpreted and implemented. The legal rights established in the Australian DDA have been viewed by some in utilitarian terms (Handley, 2001). The language is that of antidiscrimination (ADL), in the sense that discrimination must be proved in order for the case to be successful, which stands in contradistinction to the ADA's undertones of civil rights. By utilitarian terms, it is meant that this law subscribes to the idea that the government should be providing a minimal basis for disabled individuals to realise their potential as citizens, and should not stand in their way[23] (Mill, 1859). If disabled people experience discrimination, they may seek justice. However, this is different from the ADA's sweeping powers to allow any disabled individual to pursue his or her grievances in a court of law, rather than through a non-legal body such as the HREOC. Additionally, HREOC has not proved as effective a tool as had been hoped, partly because disabled people have had trouble accessing and understanding it. As an Australian interviewee noted:

> The DDA has a very high profile among people with disabilities, but they don't have as much knowledge of HREOC ... However, the only alternative is the Court process – which would put a lot of people with disabilities off especially people with intellectual disabilities and those with socio-economic concerns. This (complaints) is an affordable process (An investigation/conciliation officer for a government commission in Australia).

In the UK, the DDA is similarly well known among disabled people but sometimes misunderstood and under-utilised (Roulstone, 2003; Woodhams and Corby, 2003). The Disability Rights Commission (DRC), which was closed in 2007 and merged with the Equality and Human Rights Commission (EHRC), offered a mediation process but unfortunately the system was backlogged with complaints, so that disabled people who wanted immediate remedies for discriminatory situations

---

23   For a more detailed discussion of utilitarianism, see Chapter 1.

had to take their cases to the county courts[24] (DRC, 2005). The EHRC, and the ensuing Equality Act [2010] were thought to be a panacea for equality,[25] but as is so often the case, this commission and legislation are subject to the fiscal and ideological pressures[26] to which all other areas of government endeavour must succumb. Additionally, there have been tensions between the executive, legislative and judicial branches of government with the result that not all three fully agree on the terms and parameters of such laws.[27] Like Australia's HREOC, the EHRC faces challenges doing its work with limited resources and its decisions are not legally binding. The EHRC also works hard to raise its profile in society at large, but thus far has not succeeded in bringing about the wider cultural shifts it would like to see in terms of broader acceptance of disabled people in all spheres of public life (Roulstone, 2003). Unlike HREOC, the EHRC maintains an arms-length relationship from the government and therefore can be more critical of government policies, positions and actions or lack thereof.

Having provided the contexts in which these laws were developed, I now focus on the two main themes of the chapter. The first, in the following Section, (2.4), takes the idea of examining contexts further, and looks at how the DDA's abstract and utopic ideals impact its success, while the second, in (2.5), explores notions of how law – in its interpretation and enforcement – can harm disabled people.

(2.4) Legal Abstraction and the Importance of Context

> The letter of the law has not translated into the language of political culture. Nor has service provision broken out of (early 20th century) social welfare provision (Policy Officer for a government commission in Australia).

As already noted, the existence of legislation and rights do not guarantee equality or an end to discrimination (Doyle, 1996; Razack, 1998). This section elaborates on the debates about rights raised earlier to provide specific contextual information on how the DDA (UK) is being interpreted and enforced. I examine what kinds of resources and powers the DRC (preceding the current EHRC) was given, what impact the DDA has had, and how disability rights fit into the government's priorities and agendas.

---

24   Until its closure in 2007, the DRC had an annual limit of only 75 cases which it could afford to support.

25   See Toynbee, P (2009), "Harman's law is Labour's biggest idea for 11 years," *The Guardian*, 13 January 2009.

26   See Monaghan, K (2011), "The Equality Act is one. Will the coalition's birthday gift be to repeal key provisions?", *The Guardian*, 3 October 2011

27   For more detail on the Equality Act [2010], please see the Epilogue.

From a critical legal geographical standpoint, the DDA appears very abstract, and assumes that legislation can be applied uniformly across the UK,[28] that disabled people will take up their cases when they experience discrimination, that enforcement structures and resources are adequate, and that rights can be guaranteed regardless of the broader political climate (Doyle, 2000). I explore these assumptions, using examples from case law and interviews. Razack (1998) points out that these assumptions can help maintain existing structures of power and privilege and serve to reinforce law's power to dominate marginalised people.

The DDA, like previous ADL such as the Sex Discrimination Act (SDA), and the Race Relations Act (RRA), adheres to liberal notions of law's power to end discrimination (Doyle, 2000). It is written with the assumption that disabled people, like other citizens, are rational, independent, autonomous, self-focused individuals. With a heavy reliance on liberal notions of rights, as discussed in (2.2), the government's faith in the DDA to end discrimination ignores other factors which may impinge on the legislation's ability to do so. For example, Maria Eagle, MP, former Minister for Disabled People has said,

> This Government is committed to delivering comprehensive and enforceable civil rights for disabled people (DWP, 2004).

Yet, on another occasion, she noted that,

> Legal requirements alone won't change it... we have to win hearts and minds (speech, 2002).

Likewise, Work and Pensions Minister Anne McGuire said,

> Britain's 10 million disabled people have had to endure a legacy of exclusion, inside and outside the workplace. We have brought in legislation to help end this, but legislation can only go so far (in Mulholland, 2005).

It appears that the government wishes to use multiple strategies – legal and non-legal – to redress disability discrimination of disabled people. If, as Eagle noted, legal requirements alone cannot end discrimination, then what are the other factors that impact on the DDA's efficacy? One big cultural factor is the growing legal and political conservatism in the UK. This affects the rights, not only of disabled people, but also of women, gays and lesbians, immigrants, young people, and particularly Asians and black people in the 'war on terror' (Younge, 2005b). Rights are not acontextual, but depend on political climates within which they can be fostered and nurtured (Young, 1990). Like America and Australia, the UK is experiencing increasing conservatism in the public sphere in relation to many areas. For example, the government is cracking down on fraudulent benefit claimants,

---

28   England, Wales, Scotland and Northern Ireland all have different legal environments.

asylum seekers, and antisocial behaviour. It is a political climate in which rights are not seen as entitlements but must be balanced with responsibilities (ibid,). The general political climate is one in which disabled people feel they have to fight for their rights (RADAR, 2003). For example, the government's 2005 amendments to the DDA 1995, while ending the legislation's exemptions for small businesses and certain types of employers, came in without fanfare, and will not necessarily mean increased resources for the DRC or other enforcement mechanisms.[29]

This growing political conservatism in the UK is mirrored by similar but more drastic trends in America. This can clearly be seen in several key US Supreme Court decisions attempting to limit individual rights. These include limiting funding for, and the enforcement of, laws such as the Violence Against Women Act, the Family Medical Leave Act, and antidiscrimination protection for elderly workers. This conservatism was particularly pronounced when George W. Bush was president (2001–2008); some of his appointees, such as Justice Estrada of the DC Circuit Court, even question whether the Civil Rights Act should exist.[30] These judges have attempted to narrow the scope of anti discrimination legislation in general and to make it more difficult for disabled people to prove their disabled status, before proceeding to make an antidiscrimination claim. One interviewee noted:

> The Supreme Court has taken a more conservative bent in the last few years. There is tension between states' rights and rights of congress to enact federal legislation. There are now a number of acts in favour of states' rights, for example age discrimination legislation and the Family Medical Leave Act now weaken the federal government's powers in favour of states' rights (Director of a disability and policy organisation in the USA).

Another person concurred, noting that:

> The Supreme Court has weakened the Violence against Women act, legislation to protect elder workers, and the Family Medical Leave Act, for example (An attorney specialising in disability law).

---

29   For example, since October 2004, employment recruitment agencies are not allowed to discriminate, but these practices continue and there is no enforcement of these rules (Cottell, 2005). Other amendments include ending exemptions from Part 2 of the legislation for employers with fewer than 15 employees, and ending exemptions for fire fighters and police.

30   Estrada "refused to adequately answer numerous questions posed to him at his Judiciary Committee hearing and failed to demonstrate a commitment to the continued vigorous enforcement of critical constitutional and statutory rights in the areas of civil rights and civil liberties (www.civilrights.org). Ultimately, Estrada's nomination was blocked in the US Senate and he withdrew from the nomination process.

Likewise in Australia, conservatism, manifested through fiscal restraint and a perceived clamping down on civil liberties and rights, has undercut the efficacy of the Commonwealth disability discrimination legislation. Handley (2001) ties these results of the DDA to changes in the political and economic climates throughout what he calls the Anglo-American liberal democracies. Since the mid-1970s, and especially in the 1980s the UK, US and Australia, have been characterised by the progressive withdrawal of government from former welfare functions (Zifcak, 1994). As a Senior Policy Officer who works for an Australian government commission, noted, "we faced a 45% cut of resources in an 18 month period. Some say we're lucky to survive." In a sense, the DDA can only achieve the results it seeks with proper resources to do so, regardless of the breadth and scope of its aims. Australia has followed America's conservative trends, according to one interviewee:

> A lot of what happens in the Australia and the UK is influenced by the US. The huge economic power of America pulls others along with it. [Prime Minister] John Howard aspires to US foreign policy. There is a strong cultural imperative towards Americanisation (Head of a physical disabilities organisation).

These broader conservative trends impact political and legal cultures and can affect the material realities of disabled people. For example, as part of New Labour's policies to reduce the welfare role of the state in Britain, Tony Blair said:

> People feel it's unfair if they have to work hard, but see others getting benefits or help they're not entitled to (Blair, 2005).[31]

This culture can impact disabled people. For example, as of 2005, the DRC cited only 16 successful high-profile DDA cases since the legislation was enacted in 1995 (DRC, 2005). Some of this is directly linked to government resourcing and prioritising. During its existence, the DRC had a maximum quota of 75 cases which it can support a year (Roulstone, 2003). In a country of 60 million people, and in which disabled people face daily struggles, this is inadequate. Like America and Australia, the UK government is cutting budgets in areas which are not seen as essential to its political agenda or electoral prospects. Part of this fiscal policy was the Labour government's introduction of an Equality and Human Rights Commission which merged and supplanted the Equal Opportunities Commission, the Commission for Racial Equality and the DRC. While the previous government claimed this was an acknowledgement of how systems of oppression can cut across ability, race and gender, it also appeared to be a cost-saving measure (Adams, 2002). For example, the Northern Irish versions of these commissions

---

31   As this is taken from the website www.guardian.co.uk, there is no page number. For full reference, see Bibliography.

were merged in 2001, and according to an interviewee from the DRC, the merger was not positive for disabled people. She noted,

> Northern Ireland has an umbrella commission with remit for discrimination based on religion, race, gender, disability, and political allegiance. It came out of the Northern Ireland peace agreement and has been in existence 5 years. The dismantling of the EOC, CRE and DRC, and merging them into a single equalities body is probably going to happen. Every one of those groups will have a different set of rights. For example, in the case of an older disabled person, an employer can discriminate on the basis of age. The government is very keen to get this going but disabled people are 25 years behind the other commissions. Historically, disabled people have been told what to do by non-disabled people. The disability unit in Northern Ireland has been halved and is already beginning to see things dropping off the agenda. The whole point of any commission is, or should be, to serve the maximum number of people but the government's argument is about bureaucracy – 1 instead of 3 commissions. I think the ultimate goal is more a cost-cutting exercise than about giving redress to more people (Policy Officer for a government commission).

While all three countries are witnessing a conservative shift in the political spectrum, America still has a powerful DOJ which enforces the ADA. For example, despite the limitations to its ability to lobby for legislative change, the DOJ actively uses its clout to induce change in American society. According to one interviewee, the DOJ is said to be,

> constantly looking for cases that they believe would be precedent setting and generate a high amount of publicity (Director of a disability policy and advocacy organisation).

For example in 1996, the DOJ sued the Days Inn of America hotel group, a company that owns 1900 hotels around the world. The suits were the first to be filed by the DOJ which challenged the construction and design of buildings built after the ADA came into effect. In separate suits, the DOJ alleged that five hotels, in Indiana, Illinois, South Dakota, Kentucky, and California, were constructed after the enactment of the ADA, which required builders to comply with specific architectural guidelines ensuring that 'persons with disabilities' could gain access to the facilities. Other examples of cases included high-profile settlements with testing agencies for the Scholastic Aptitude Test, the Law School Admissions Test, and the Graduate Record Examination, such as Kaplan and Princeton, as well as Super Shuttle, an airport-hotel shuttle company (Manager of a disability technical assistance centre, in interview).

Similar examples are difficult to find in the UK, partly because of a lack of resources and partly because the enforcing agency, the EHRC, is not as powerful as the DOJ.[32] As noted earlier in the comments by the Minister for Disabled People, the government emphasises the rights of disabled people rather than discussing the resources or powers needed to create equality. While utopic liberal ideals of equality and rights make for captivating electoral slogans, they do not always translate into concrete action which makes society more inclusive of disabled people. As one interviewee noted:

> There is a dissonance between the principles of social equity – which are in all university mission statements – and the way that gets played out in practice when there are scarce resources and competing demands. The broad brush of social equity goals get lots in the day to day operations of different levels of management in the bureaucracy, with different levels of understanding, and different aims and goals (Policy Officer, Australian government commission).

In the next section I examine how the legislation itself can harm disabled people.

(2.5) How Law 'Harms': Disabling Aspects of the DDA

Building on the last section's arguments that the DDA alone is not enough to end discrimination against disabled people, this section takes the discussion one step further: how the exercise of their rights under the DDA can actually cause harm to disabled people and how these legal processes can themselves be discriminatory, demoralising and dehumanising (Roulstone, 2003). Using interview material and examples from case law, I indicate that the DDA's enforcement mechanisms, like its American and Australian counterparts, can have negative impacts on disabled people. This is because of the narrow, medicalised assumptions of the legislation and because of the continuing ableist attitudes and discourses in courts, tribunals and other legal processes (ibid.).

The DDA (UK) is seen as ineffectual, according to some disability activists and theorists because of its individual-medical view of disability (Oliver and Barnes, 1998; Roulstone, 2003). The individual-medical model of disability casts a person's impairment – either in terms of injury, a missing limb or function, or mental or physical illness – as the cause of their disability (ibid.). The social model of disability, while taking into account the role impairment can play in shaping one's life, sees society as disabling the individual through its inaccessible environment and insensitive cultural norms and biases (Gleeson, 1999). While this particular criticism has been launched at the UK DDA, the Australian and American laws are premised on not dissimilar models of disability, which presume societal structures as given, and which must be adjusted to reasonably accommodate disabled people.

---

32   One interviewee notes that despite the DOJ's extensive powers and large budget, because of the extent of disability discrimination in the USA, the DOJ is very 'backlogged and overwhelmed.'

In relating this to Young's (1990) critique of the distributive paradigm of justice mentioned in Chapter 1, the DDA can be seen as attempting to distribute more social and legal goods to disabled people – such as housing, jobs, income, or access to buildings. Furthermore, we can use Young's argument that dominant social groups, such as able-bodied people, work to maintain hierarchical structures of power (ibid.). Disabled people have to fight for their rights by proving – within the confines of the legal system – that they experienced discrimination. Roulstone (2003:122) notes that in order to critically unpick the DDA we must examine and disentangle the manner in which "disability, offence and harm are constructed and applied." He notes that in spite of some high profile DDA case law (e.g. Kirker v. British Sugar and Hedley v. Aldi Supermarkets) and in spite of the large number of cases presented to the DRC,[33] the legislation itself contains aspects which limit its effectiveness (see Figure 2.1).

As mentioned in (2.3), the DDA can be used by a disabled person when they can prove they have been treated 'less favourably' and can show that this relates to a 'disability' as defined by the Act.[34] However, as Roulstone (2003:122) notes, "further hurdles need to be jumped." One condition is that the disability must be 'substantial' and 'long term,' concepts which are taken from the Chronically Sick and Disabled Persons Act 1970 (Gooding, 1994; Woodhams and Corby, 2003). The DDA's conception of disability is limited. For example, from 1995 to 2005, it failed to include people who may be stigmatised for unseen impairment such as people living with HIV, or those with perceived impairment such as people with facial disfigurements.[35]

Another limiting feature of the DDA is its insistence that disabled people specify what type of 'disability' they have. For example, in Goodwin v. Patent Office (see Figure 2.1) where the complainant was a worker with schizophrenia, which falls outside the given legal categories, his 'condition' had to be slotted into several different categories including 'difficulties with coordination' and 'lapses in memory and concentration' (Roulstone, 2003). As Oliver and Barnes (1998) and Roulstone (2003) point out, the labelling and categorising of various and multi-faceted relationships between impairment and social conditions into legally abstract categories is ineffective and not necessarily inclusive. The use of these categories in the DDA is evidence of able-bodied privilege in that the key definitions and power to label have been created with inadequate acknowledgement that disabled people's realities cannot be limited to categories.

---

33   Only a small proportion of applications to the DRC were taken up, detracting from the Act's impact.

34   The DDA's definition of a disabled person is someone with "a physical or mental impairment which has a substantial and long-term adverse effect on his ability to carry out normal day-to-day activities." (HMSO, 1995).

35   The 2005 Amendment to the DDA includes unseen impairment, but there is no case law to date, and this amendment came ten years after the original Act, and is due to the work of pressure groups such as RADAR. For more detail on this Amendment, see Chapter 7.

Goodwin v. Patent Office

The applicant had paranoid schizophrenia and was dismissed by his employer following complaints by colleagues. He imagined that other employees could access his thoughts, and misinterpreted words and actions of colleagues due to paranoia. He also left the office frequently due to auditory hallucinations. However, he could care for himself at home, dealing with shopping, cooking and personal hygiene without help. The industrial tribunal held that the adverse effect of the impairment on normal day-to-day activities was not substantial, since he was able to perform his domestic activities without assistance and to carry out his work to a satisfactory standard. He then appealed this ruling, and the Employment Appeal Tribunal found that he was 'disabled'. The fact that Mr. Goodwin was able to cope at home did not mean that he was not covered by the legal definition of a disabled person for the purposes of the Act.

The EAT gave detailed guidance generally on the proper approach for determining whether a person is) "The focus of attention required by the Act is on the things that the applicant either cannot do or can only do with difficulty, rather than on the things that the person can do" and, (2) "The tribunal may, where the applicant still claims to be *suffering* from the same degree of impairment as at the time of the events complained of, take into account how the applicant disabled. Two interesting quotes, which are oft-cited, appeared in the appeal ruling: (1appears to the tribunal to 'manage'."

Kirker v. British Sugar

Nick Kirker won what has been called a "landmark" Law Lords ruling in 2003, giving him the right to challenge a negative reference provided by his ex-employer (source: BBC News, 2003). Mr Kirker had not been able to find work since 1997 when he successfully sued British Sugar for disability discrimination. The court ruled that Nick Kirker and other disabled people could legally challenge the discrimination or victimisation they claim they faced from former employers. Mr Kirker, who had visual impairment, challenged a reference by his former employers in which they indicated that they would not employ him again. He felt that the reference was unfair and inaccurate and noted that his former employers were biased against him out of resentment as he had taken successful taken his case against them to a disability discrimination employment tribunal. In March 1997 Nick Kirker was made redundant from his job as a chemist but he took British Sugar to a disability discrimination employment tribunal. He won his case and was awarded £103,146, the highest award under Part 2 at the time. In August 1999 he applied for a job as a warehouse assistant but he claimed that he was not given the job on account of a negative reference given by his former employer, British Sugar.

(Note: there were two separate cases: (1) he was fired from British Sugar and he successfully challenged that dismissal, (2) he successfully sued them for providing negative references to other would-be employers].

Source: www.employmentappeals.gov.uk

Figure 2.1   Key disability discrimination cases in the UK

Beyond the definitional limitations of the DDA, its exercise and procedural aspects can also cause difficulty for disabled people. As Roulstone (2003:123) notes:

> Legal power attaches to lawyers and tribunal members' positions as arbiters of what is or is not an impairment, a disability, substantial effects, long term and so on.

An acknowledgement of this legal power is important, as it helps to demystify the law and to destabilise claims of legal impartiality (Gooding, 1994; Razack, 1998; Young, 1990). As Blomley et al. (2001) point out, notions of law's neutrality in classical jurisprudence are highly contested by Marxists, feminists, anti-racists and post-structuralists. Feminist and anti-racist legal theorists have argued that legal bias illustrates the conservative and patriarchal underpinnings of canonical law (Matsuda et al., 1993; Razack, 1998). For example, until the Domestic Violence and Matrimonial Proceedings Act (1975) those working in the legal system – including police and judges – constructed domestic violence as a private issue beyond the law's reach, and women's testimony about their own cases were sometimes ruled to be *ultra vires* (Pannick, 1987).[36]

A few key cases illustrate the above issues such as definitional disputes over disability, legal bias at work in the justice system, and 'harm' being inflicted on disabled people. In Goodwin v. Patent Office, for example, the respondent's lawyers succeeded in persuading the tribunal that Mr. Goodwin's day-to-day work activities were not substantially affected by his paranoid schizophrenia about which the employer was aware. This is similar to the Toyota v. Williams case in the USA in which a woman who developed a repetitive strain injury from work on an assembly line was not deemed disabled because the injury did not affect her day to day activities (see Figure 2.2). As one American interviewee noted:

> Most people still think it's okay to discriminate against people with disabilities because there are not that many of them and disability is ingrained in the public's minds as incompetence ... for example in the case of Toyota vs. Williams – the woman was damaged by her employer performing manual tasks – as a major life activity. Yet they refused to acknowledge her disability and she went on to worker's compensation. They would rather pay her not to work than acknowledge her disability. (An attorney who specialises in disability law).

According to UK law reports, in hearings where the notion of day-to-day activities is contested, complainants are told to list the things they cannot do as evidence for the trial (Woodhams and Corby, 2003). In the case of Goodwin, (see Figure 2.1) the tribunal emphasised that the DDA focuses on what an applicant cannot do, emphasising the degree to which they stray from 'normality.' The tribunal also

---

36   For example the evidence of Greenham Common protesters was ruled to be *ultra vires*.

Toyota Motors v. Williams

This case saw the plaintiff lose the case, and garnered much attention as a setback for the disability movement in America. Williams, the plaintiff, had a repetitive strain injury due to her work on the assembly line, but the Supreme Court justices did not recognise her as disabled because this injury did not affect her "normal day-to-day activities." This was despite the fact that she developed the injury due to her job. However, the court interpreted the language of "normal day-to-day activity" as excluding manual labour, and thus disqualified her as being disabled. As a result, she did not receive the accommodation she requested. This case has been viewed as a victory for employers because it narrowed the definition of disability under the ADA. The unanimous decision, authored by Justice Sandra Day O'Connor, said that to qualify under the law, Williams' disability must extend outside the workplace to affect activities of daily life, despite the fact that the injury was caused by her workplace, and impedes her ability to do her job. There have also been other similar cases, and there is talk of strengthening disabled peoples' rights by asking Congress for new legislation to reinforce the rights set out in the ADA.

US Airways v. Barnett

Robert Barnett injured his back while working in a cargo-handling position at the US Airways. He requested a transfer to a less physically demanding position in the mailroom and was temporarily placed there. He then learned that at least two employees senior to him intended to bid for the same job. He asked the airline to accommodate him by making an exception that would allow him to stay in the mailroom. The airline decided not to make an exception, and Barnett lost his job. The judges, for the first time, examined the meaning of "reasonable accommodation" under the provisions of the ADA governing private employers. In this case the Supreme Court was asked whether employers must make a requested accommodation even if it conflicts with an employer-established seniority system. They voted 5-4 against Barnett's case, arguing that making an exception to a seniority rule did not constitute a reasonable accommodation, as it had nothing to do with his disability-related needs. As a general rule, the Court held, employers do not have to disturb an established seniority system to accommodate the needs of a disabled worker. Justice Sandra Day O'Connor provided dissenting comment saying that employers' seniority systems should only be strictly adhered to if they are legally enforceable. This proved to be very contentious, with advocates saying this decision means disabled workers seeking reasonable accommodations from their employers will face a higher hurdle in showing the reasonableness of their request. There were repeated calls for Congress to amend the Act and neutralise the Supreme Court's rulings.

Source: www.eeoc.gov

Figure 2.2   Key disability discrimination cases in America

used the patronising language of 'suffering' to describe Mr. Goodwin's situation. As Roulstone (2003) notes, there is no other form of UK anti-discrimination law in which justice is based on people listing their limitations and shortfalls. He says (2003:123):

> It is not natural or uplifting for a person to have to dwell publicly upon the minutiae of what they cannot do.

This requirement is not unique. As mentioned in (2.2), the government has been reducing incapacity benefits, and claimants have to provide more evidence of their impairments. The difference is that in the case of the DDA, the tribunal findings are made public and disabled complainants undergo verbal questioning. As Morris (1991), Razack (1998) and Roulstone (2003) note, the impacts of legal procedures on people marginalised by legal systems, including disabled people, requires further exploration. Roulstone (2003:123) says:

> It is worth reflecting on whether the legal processes of employment tribunals, and the DDA themselves might disable significantly or long-term the integrity, and confidence of disabled complainants, whatever the outcome.

Likewise an interviewee noted that the process of taking a case forward can adversely affect disabled people:

> You only have to look at the Sex Discrimination Act, and the Race Relations Act ... they're similar ... And how many people have been taken, how many universities have been taken to court under race relations or sex discrimination? Hardly any. Because what happens is, the onus is on the individual, they have to go through a massive big rigmarole in terms of procedure, it has to be seen that they've pursued everything within the internal disciplinary agreements, procedures, etc. The onus is on the student to prove they have a disability, and this alone can have negative impacts on them (Team leader for a funding council for higher education).

Roulstone (2003) notes that while similar claims of racism and sexism in the justice system have been made, there are 'uniquely disablist factors' in the processes of scrutinising and contesting impairment and limitations to daily life activities. Roulstone (2003) and Razack (1998) feel that an acknowledgement of the potential for judicial harm is important, where the conception, interpretation and enforcement procedures of the law can themselves be harmful and impact the confidence and integrity of disabled people who employ these legal tools. As an interviewee noted:

> Giving testimony – what people have gone through is so earth shattering. Because of this, one third of people who are disabled and face discrimination

don't do anything about it, because of stigma ... they have to face the patronising attitudes of legal experts and the language they use. There is a still a general taboo about disability. People feel pity towards disabled people and think that 'disabled people cannot do it for themselves' [37] (Policy officer for a government commission).

Another interviewee, a policy officer in London, said:

It's quite a difficult thing. I mean I think disabled people are disadvantaged by the fact that the law, there is nothing there to help you comply. You have to pay for yourself, it all adds up. It's a very disempowering approach (A local authority access officer and architect).

Likewise a policy officer in Sydney concurred:

It's hard to get people to lodge complaints; the energy and emotional costs to them. They are nervous about the public aspect and people with disabilities are inevitably the weaker at the table. For example, it is difficult, in higher education, for students to complain – it is time consuming, difficult, and face they chances of retaliation (A senior policy officer for a government commission in Australia).

The case of Kirker v. British Sugar shows that retaliation against disabled people who complain does indeed occur. This case also illustrates the persistence of ableist biases in society and the legal system. Even after having lost their wrongful dismissal on the grounds of disability discrimination case to Mr Kirker, and having had to pay £103,146, British Sugar still gave negative references on his behalf to prospective employers. Such subtle but powerful forms of discrimination would be difficult to catch were it not for the persistence of Mr. Goodwin. This case is astonishing, but also indicates the extent of ill feeling these employers had towards this individual, and their incorrect assumption that they might get away with it. This case reveals similar issues to Daghlian v. Australian Postal Corporation, in which an organisation was successfully challenged by a disabled person in the case of Heath v. Adelaide General Post Office, only to perform further discriminatory acts towards Sarah Daghlian. Additionally, as in the case of Toyota v. Williams, these cases highlight an implicit assumption that disabled people are better off not working, and being paid to not work, if necessary (see Figure 2.3).

As a result of these legal hurdles, the success rate for disabled people contesting disability discrimination is low (DRC, 2005; Meager et al., 1999; Woodhams and Corby, 2003; Roulstone, 2003). Of the 5662 cases lodged from 1996 to 2000, 457 (8 per cent) proceeded to tribunal and in 105 cases (1.9 per cent), disabled people were able to prove that they had been treated 'less favourably' (Roulstone, 2003).

37   This person provided this statistic based on the number of complaints lodged with her office.

Cocks v. The Queensland Government

Disabled people could not access the convention centre by stairs. People with mobility impairment were either excluded from entering the building, or made to enter via the "tradesmen entrance" down the alley past the rubbish bins, and in the back door. The Convention Centre case was one of the landmark decisions in terms of access to premises. It confirmed that the Building Code of Australia (1992 edition) was a discriminatory document, and brought about an immediate update of the code. The decision was a significant one because it focussed the building industry's attention on lawful requirements for equitable access to public premises. It also influenced the DDA to such an extent as to allow for an Access to Premises Standard to be developed, something that was not envisaged in the original legislation.

Daghlian v Australian Postal Corporation

Sarah Daghlian was in employment in the postal service, a public service corporation. The service later had all premises remodelled, and introduced a chair policy banning stools introduced for counter employees. Daghlian was removed from employment at the counter under the auspices of the chair policy. She was then placed on long service leave, and upon expiration of that leave her employment was terminated. The Federal court case was to determine (1) whether enforcement of the chair policy was unreasonable in the particular circumstances of that disabled employee, (2) whether the employee was unable to carry out inherent requirements of the particular post, (3) whether the auxiliary aid in the form of a chair or stool required by the disabled employee and not required by other employees imposed unjustifiable hardship on employer. Like Williams in the US Toyota v. Williams case, Daghlian was injured on the job, while lifting heavy parcels. The Court found that terminating her employment was indeed disability discrimination. This case is interesting because it highlights the different approach of Australian and American high courts. In Toyota, removing Williams from the post was not seen to be discriminatory, but in Daghlian it was. This case also highlights the fact that even after the Jeff Heath case against Adelaide General Post Office, cited earlier, Australian Post as an organisation does not seem to have implemented organisation-wide polices and procedures for inclusion of disabled employees and customers. After having lost the Heath case, one would think the corporation would devise strategies to avoid being sued again.

Source: www.hreoc.gov.au

Figure 2.3   Key disability discrimination cases in Australia

A large majority of unsuccessful cases – 80.5 per cent – failed, not because disabled complainants were unable to establish 'less favourable treatment', but on the grounds of legal exclusions.[38]

---

38 For example, "13.1% of case submissions fail because the applicant is deemed 'not disabled' for the purposes of the Act...12.8% fail because their impairment is 'not substantial', while 11% fail because their claim is 'out of time' (Roulstone, 2003:124). A total of 80.5% cases to date have failed on grounds of legal exclusions (Roulstone, 2003).

Another factor which limits the ability of disabled people to exercise their DDA rights is the issue of legal representation and the associated costs. As Roulstone (2003) notes, it is possible that there is a relationship between the success rates of disabled people's cases and their ability to present themselves as credible. For example, one study indicated that the success rate for barristers working on disability discrimination cases was 40 per cent, while it was 32 per cent for solicitors, 25 per cent for trade union lawyers, 17 per cent for CAB legal workers, and only 15 per cent for self-represented cases (Income Data Services, in Roulstone, 2003). Additionally, 21 per cent of cases reached tribunal with disabled people representing themselves, and there is clearly a lack of legal aid funding to provide better representation (ibid.). As Razack (1998) notes, there are performative aspects to courtroom proceedings, and ultimately outsiders like disabled people, women and aboriginals have to convince judges of their adherence to the law's normative values and terms of reference. She says that storytelling is,

> An interrogation of how courts come to convert information into fact, how judges, juries and lawyers come to 'objectively' know the truth. 'Those whose stories are believed have the power to create fact.'[39]

As argued by numerous theorists, the ways in which the law defines, labels and categorises disabled people actually shapes their realities, how they are perceived in courts and tribunals, and whether or not their claims of discrimination are seen to be credible (Barnes, 1991; Benjamin, 2002; Foucault, 1979; Gooding, 1994; Gleeson, 1999; hooks, 1994; Razack, 1998). This demonstrates law's power to influence people's lives. Taken together with the arguments made in (2.4) that the DDA is too abstract, and its legal contexts too weak – in terms of resources, enforcement, and the political will to create inclusion – disabled people might find these analyses disillusioning. What are the ways in which inclusion for disabled people can be created? The final section of this chapter summarises its key findings and lays the groundwork for the remaining chapters of this thesis.

(2.6) Conclusions

> Although the DDA has not transformed the life experience of all people with disabilities, it has provided a mechanism with which to fight some forms of injustice (Disability rights advocate in Australia, in interview).

In Australia, the effectiveness of the DDA appears to be at the whim of an increasingly hostile commonwealth government (Handley, 2001). In America, while the government has not introduced measures to limit the ADA, or reduce

---

39   Kim Lane Scheppele (1989:2079) quoted in Razack (1998:37). The idea of legal storytelling is further explained in Chapter 6.

the DOJ's resources, its appointment of conservative judges, and perpetuation of an anti-civil rights ideology threatens the progress of disability rights (ADA policy officer, in interview). While the DDA is relatively new, and legal strategies for ending disability discrimination are still developing, the UK can learn from the experiences of America and Australia. Because of the impact of conservative legal and political discourses in those two countries, disabled Americans and Australians are pleased that they have disability discrimination laws, but are also aware that CRL is not enough to improve their lives.

The research conducted for this book reinforces the arguments made in (2.1) that law is idealistic, removed from everyday realities of disabled people, and that law itself has led to more categorisation, stigmatisation and 'othering.' Reasonable adjustments and accommodations emanating from the DDA, and its Australian and American counterparts, are related to over-arching medicalised notions of disability, and the rights granted to disabled people are constantly under threat in the courts of all three countries.

The DDA (UK) only envisages discrimination as something that happens to, or is done to, disabled people, but it does not examine the role of able-bodied people in creating and maintaining their privileges. As indicated in (3.5), the onus is on individuals to prove that they are disabled, and that they faced discrimination. No similar onus rests on defendants to prove that they did not discriminate, and the actions of employers, goods and service providers, and educational institutions are enframed in a language of reasonableness, a concept which is vague and nebulous, unlike the DDA's definition of disability, which is grounded in liberal medical roots. The impacts of this medicalised definition are further examined in the context of service provision in Chapter 3, and in the context of higher education, in Chapters 4, 5 and 6.

Chapter 3
Deconstructing Liberal Discourses in Service Provision

(3.0) Introduction

As Chapter 1 intimated, in liberal conceptions of law, equality is thought to be achieved by ensuring the legal rights of all members of society, with laws created and enforced by neutral and independent policy makers and practitioners (Ramsay, 1997; Razack, 1998). In reality, liberal conceptions of legal equality prove much more complex, and are based on liberal individualist theories (Young, 1990). For example, the liberal idea that the state should essentially leave individuals to do as they please limits the state's ability to change attitudes towards disabled people (ibid.).

Given its liberal legal foundations, how does the government envisage the inclusion of disabled people in society through the context of the DDA? The DRC, a government-appointed independent body charged with advising the government and the general public on disability discrimination issues, and with supporting benchmark cases and arbitrating and mediating settlements from 1997 to 2007, said,

> Legislation is no guarantor of social change; nor can the law operate in isolation from other social and economic forces. The causes of social change are complex, the reasons for legislative impotence various. To stand a chance of success, the law must be adequately framed; the courts and tribunals must be sympathetic to its general interpretation. The alternative is erosion and obsolescence (DRC, 2002a:1).

What are these social and economic forces? The DDA, which seeks to end discrimination against disabled people, is the product of a legal system premised on able-bodied norms and values, but which claims to be abstract, neutral and universal (Gooding, 1994; Woodhams and Corby, 2003). The term 'able-bodied' is employed both because it is considered acceptable by disabled people, and to draw attention to the fact that social and physical environments are primarily designed around the needs of the majority of people (Morris, 1991). Many people are unaware of the social privileges that go along with being able-bodied, and the use of this language is a deliberate attempt to raise consciousness of these privileges.

In equality struggles, discourses and language are important, as they reveal the inner workings and meanings of policies and documents, and this is recognised in many areas including law and geography (Blomley et al., 2001), education (Razack, 1998), and planning (Hillier, 1993). Language is significant, not just for the sake of political correctness but also because the language we use reveals our underlying values (Langton-Lockton, 2000). Given this significance, this chapter analyses not only what the law says, but also its underlying assumptions and values.

Part 3 of the DDA broadly requires providers of goods, services or facilities in the UK – identified as 'service providers' in the legislation – to make their services accessible. This chapter examines Part 3 of the DDA and illustrates its liberal approach to the inclusion of disabled people, building on what others have written about it, and on testimony from research interviews (Cooper & Vernon, 1996; Gooding, 1994). This research illustrates how Part 3 of the DDA is being interpreted and acted upon by service providers in the UK and I highlight the connections between the theoretical underpinnings of the DDA and the material realities of what service providers are doing.

(3.1) Part 3 of the DDA

Part 3 of the DDA places duties on all organisations and businesses that provide goods, services or facilities to the public (except in certain sectors such as education and transport, to which Parts 4 and 5 pertain, respectively). This includes a broad range of service providers including shops, restaurants, petrol stations, and other businesses, as well as government departments, local authorities, doctors' surgeries, public libraries, museums, galleries and community centres. Part 3 has three phases: (1) since 2 December 1996 it has been unlawful for service providers to treat disabled people less favourably than others for a reason related to their 'disability',[1] (2) since 1 October 1999, service providers have had to make 'reasonable adjustments' for disabled people in the way they provide their services, and (3) since 1 October 2004, service providers have had to make 'reasonable adjustments' to the physical features of their premises to overcome physical barriers to access (HMSO, 1995).

A more thorough understanding of Part 3 requires an examination of its definitions and descriptions of discrimination. Part 3 of the DDA makes it unlawful for a service provider to discriminate against a disabled person in three ways: (1) by refusing to provide (or deliberately not providing) any service which it provides (or is prepared to provide) to members of the public, (2) in the standard of service which it provides to the disabled person or the manner in which it provides the service, (3) in the terms on which it provides a service to the disabled person. It is also unlawful for a service provider to discriminate by failing to comply with any duty imposed on it by section 21 (a duty to make 'reasonable adjustments'), where the effect of that failure would make it impossible or unreasonably difficult for the disabled person to make use of any such service (HMSO, 1995).

---

1   As noted in Chapter 1, the DDA employs medicalised definitions of disability. According to the social model of disability, the term 'disability', as employed in the legislation, should be replaced with 'impairment'.

The DDA states that discrimination against a disabled person occurs in two possible ways. One would be when a service provider treats a disabled person less favourably – for a reason relating to the disabled person's disability – than it treats (or would treat) others to whom that reason does not (or would not) apply, and cannot show that the treatment is justified. The other way in which discrimination occurs is when a service provider fails to comply with a duty imposed on it by section 21 of the DDA (a duty to make 'reasonable adjustments') in relation to the disabled person, and cannot show that the failure is justified (HMSO, 1995).

There are six possible conditions under which it is possible to justify unequal treatment: (1) if the less favourable treatment is necessary in order not to endanger the health or safety of any person, (2) if the disabled person is incapable of entering into a contract,[2] (3) where refusing a service is necessary because the service provider would otherwise be unable to provide the service to members of the public, (4) where a person has been treated less favourably in the standard, manner or terms on which the service is provided because this is necessary in order to provide the service either to the disabled person or to other members of the public, (5) where a difference in the terms in which the service is provided to disabled people is due to a greater cost to the service provider, and (6) where refusing service to a disabled person is necessary to protect the fundamental nature of the business or service (HMSO, 1995).

These justifications for unequal treatment have been criticised by some because they say it effectively means that service providers can discriminate where reasonably justified (Gooding, 2000). Doyle (2000), a lawyer and legal scholar, says that the defence of justification weakens the DDA in that it could give service providers a loophole for excusing their unequal treatment of disabled people. Indeed, this has played out in Part 2 litigation, and in one case, the employer was able to justify their discrimination by claiming that they had limited knowledge of the employee's 'disability' at the time (Woodhams and Corby, 2003). Case law to date indicates that the service providers' justifications have largely been taken at face value[3] (Doyle, 2000). To counter this, the Act places anticipatory duties on service providers by requiring that 'reasonable adjustments' be made in relation to disabled people at large. It is not enough for a service provider to wait until a disabled person has difficulty accessing their service to rectify the lack of access. Service providers are encouraged to think about the accessibility of their goods, facilities or services to disabled people generally, and to plan to meet a diverse range of needs.

Service providers are expected to make adjustments as long as they are reasonable. Adjustments are not required when they are unreasonable. However, unlike Part 2, there are no indications in the Act itself as to what factors will

---

2   Some would see this as a patronising assumption, i.e. the idea that some disabled people are incapable of understanding what a contract is, or being able to agree to one (Oliver and Barnes, 1998).

3   For more detail on case law, see Chapter 2.

be taken into account in determining reasonableness. Reasonableness itself is an ambiguous legal concept. For example, the term reasonable is being removed from the criminal code in certain jurisdictions such as Canada (Moran, 2003). The Code of Practice – a non-binding guideline which accompanies the DDA – suggests factors which should be considered when defining 'reasonableness,' including (1) whether taking any particular steps would be effective in overcoming the difficulty that disabled people face in accessing the services in question, (2) the extent to which it is practicable for the service provider to take the steps, (3) the financial and other costs of making the adjustment, (4) the extent of any disruption these steps would cause, (5) the extent of the service provider's financial and other resources, (6) the amount of any resources already spent on making adjustments, and (7) the availability of financial or other assistance (DRC, 2002b). Having outlined the basic tenets of Part 3, I will examine what others have said about it in the next section.

(3.2) Analysing the DDA's Discourses

In examining the liberal theoretical perspectives which underpin the DDA, it is important to consider what definition of disability is used in the Act. The government's publication entitled *What Service Providers Need To Know,* defines a disabled person as "somebody whose disability makes it difficult for them to carry out normal day-to-day activities" (DfEE, 2001:3). The complete definition, as outlined in the legislation, says (HMSO, 1995:1) that,

> A person has a disability for the purposes of this Act if he [sic] has a physical or mental impairment which has a substantial and long-term adverse effect on his [sic] ability to carry out normal day-to-day activities.

Disability activists and scholars see this as problematic in that it frames the individual person as inherently 'disabled' rather than being disabled by social barriers and biases in the built environment. This definition also uses the term 'disabled' to describe individuals, rather than describing them as having 'impairments,' a term which some theorists prefer (Oliver and Barnes, 1998; Reinders, 2000). In contrast to this definition of disability, which some would describe as a medical one, the World Health Organization's definition of disability focuses on the relation between individual capacities and sociocultural environments (Reinders, 2000). Finally, the DDA's definition of disability uses the term 'normal' to describe people's day-to-day activities. As mentioned in Chapter 1, many critics of assert that the normalisation and assimilation of difference – in relation to identity – are key tenets of liberal doctrine (Ramsay, 1997).

The issue of normalisation is not just a theoretical one, but also presents problems in practice. For example, in the Part 2 case Quinlan v. B and Q plc, the applicant, who had undergone open-heart surgery, was dismissed from his job in a garden centre, as he was unable to lift heavy objects (Woodhams and Corby, 2003). However the Employment Tribunal found that despite this impairment, he did not qualify as being disabled under the Act, because lifting heavy objects is not considered a normal day-to-day activity (ibid.). This means that although disabled people may be impeded from carrying out their particular vocation, they are not considered disabled because they do not fit the narrow legal category set out by the Act. As Woodhams and Corby (2003:168) note, "this decision means that a concert pianist who is unable to play the piano would not be held to be disabled by his impairment if he were able to continue day-to-day activities" since playing the piano is not considered a normal day-to-day activity, even though this person's livelihood depended on it. This is confirmed by the decision of the Employment Tribunal in the case of Goodwin v. The Patent Office (as noted in Chapter 3):

> What can be said is that the inquiry is not focused on a particular or special set of circumstances. Thus it is not directed to the person's own particular circumstances, either at work or home. The fact that a person cannot demonstrate a particular skill, such as playing the piano, is not an issue before the tribunal, even if it is considering a claim by a musician (Woodhams and Corby, 2003:168).

This demonstrates that the Act does not necessarily concern itself with the particular facts and complexities of actual disabled peoples' lives, but rather is centred on a theoretical disabled person, who falls within liberal conceptions of abstract legal subjects. Liberalism claims to put the interests of individuals before those of society (as opposed to communitarianism, for example), but does not really grapple with individual difference so much as it focuses on the individual as a singular unit – neutral, abstract, and to be treated the same as all others. Thus, disabled people are stigmatised by the DDA, not only through the essentialising of disability through its definitions, but also because the law ignores their real and actual needs by virtue of those definitions.

The DDA's definition of disability attempts to apply the law according to narrowly-defined categories, in which disabled peoples' abilities are measured against a standardised list of of 'normal' abilities, and this ignores the contextual issues and particularities of individuals. Paradoxically, the DDA employs an individual case-by-case approach for redressing discrimination, yet does not allow or account for individual needs, abilities and differences. The Goodwin v. Patent Office case is also an example of how the legal system – in this case a tribunal – interprets the law in a way which does not advance equality for disabled people. This proves that even the best legislation, implemented and enforced in a legal system which is not committed to equality for disabled people, is not enough to create equality.

Because of its view of disability residing in individuals, and its use of comparisons to 'normality,' the Act assumes a medical model of disability (Gooding, 1994; Woodhams and Corby, 2003). As noted in Chapter 1, a medical model assumes that a lack of, or altered ability in a particular functional area of the body or mind makes an individual disabled. Under this definition, disability is conceptualised as an individual phenomenon. Disabled people, seen as different from the 'norm,' are unfortunate, sad, and lacking in some area of their life (Oliver, 1990). As Gooding (1994) and Doyle (2000) have pointed out, the DDA's medical view of disability, with its linking of impairment to the ability to carry out day-to-day activities, does not allow for social or environmental factors which may exaggerate or alleviate the impacts of impairment.

As alternatives to the medical model, there are multiple theoretical conceptions of disability under the broad heading of 'social model' which seek to define disability as a disadvantage which is deeply rooted in the relations between people and the social and built environments (Barnes, 1991). As Woodhams and Corby (2003:164) note,

> The suggested essence of disability, then, rests on the social and economic consequences of being different from the majority. As a result, the territory of 'correction' becomes society and the environment, rather than the person with the impairment.

The language of 'territory' is interesting. It suggests that overcoming barriers to access is not about 'curing' individuals but about changing the environments which cause the barriers in the first place. I would add to this by suggesting that the disadvantages of disability occur in the physical and social environments that are built by humans, that discrimination has materiality, and that space and context matter. To use a medical model of disability is to discount the effects of able-bodied norms in the material realities of the world (Pue, 1990).

The definition of disability in the DDA is central to the Act's statutory purpose in that those who claim they have been discriminated against must first prove that they are indeed disabled. As such, it is contentious in two opposing ways. For those wishing to restrict the number of people claiming rights under the DDA, the definition serves to narrow the category of 'disabled' and therefore limit damages under civil lawsuits. Indeed, this kind of 'narrowing' has already occurred in relation to the ADA in America, for example in the case of Toyota v. Williams, as described in Chapter 2.

Conversely, disabled people argue that the definition should be as broad as possible so as to include the greatest number of individuals and therefore target discrimination in as many cases as possible (Gooding, 1994). The conflict between these two perspectives has been so great that the Disability Rights Task Force (DRTF), a government-appointed body established to review the DDA's progress, regarded the issues surrounding the definition of disability the most difficult ones to consider (DRTF, 1999). For example, the definition of disability does not include those who are infected with HIV, until the symptoms of illness are exhibited. It could take years for such symptoms to manifest after people are diagnosed with HIV.

Yet, such individuals may and do face discrimination due to their HIV status regardless of whether the illness is visibly manifested or not.[4]

Additionally, having to prove one is disabled by using tools of measurement and medical assessment to demonstrate a lack of certain 'functions' as they relate to particular activities can be seen as a negative view of disability (Roulstone, 2003). How must it feel for a disabled person to experience discrimination, and then have to prove, in a county court or an employment tribunal, that they cannot do something, or have a perceived inability or lack of function in a particular area? What might be the psychological and emotional effects on disabled people in having to compare themselves to those who 'can' and label themselves as 'other,' a deviation from this normalised standard – the person who can do everything? What might this do for their sense of self-worth?

There is much literature on the psychological toll taken by these legal processes on disabled people and therefore many disabled people do not take up legal cases to redress discrimination[5] (Handley, 2001; Roulstone, 2003; Woodhams and Corby, 2003). As the Minister for Disabled People noted:

> There are 8.5 million disabled people. We don't want them all to have to go to court to get their rights. It will cost them a lot of money, it will drive them up the wall, and we know that a lot of them don't go that far, they will just put up with it (Eagle, 2002).

The DDA's definition does include emotional, mental, physical and/or sensory impairment, and is therefore broader than the ways in which disability was previously defined, for example, in the Disabled Persons (Employment) Act 1944. There is limited recognition of social factors leading to discrimination, for example in the case of discrimination against those with facial disfigurements even where these impairments do not necessarily have medical effects (Woodhams and Corby, 2003).

As noted in Chapter 2, many writers argue that the UK definition of disability is more restrictive than that of other jurisdictions such as Australia and the USA (Doyle, 2000). The ADA and the Australian DDA, for example, include coverage for individuals who may be perceived as disabled although they are not, and for those who face discrimination due to their association with disabled people (HREOC, 2003; US DOJ, 1992). The DDA (UK) also excludes coverage for those discriminated against due to people's fears of contracting illness, and this includes those with genetic disposition to certain illnesses, for example (Woodhams and Corby, 2003).

Like the Sex Discrimination Act (1975) and the Race Relations Act (1976), the government has opted for an apparently 'obvious' biologically-determined as

---

4 As of January 2005, HIV is now included in the Act's definition of disability (but not at the time interviews of service providers were conducted).

5 For further discussion of disabled peoples' experiences of taking up legal cases, see Chapter 2.

opposed to socially-constructed criteria to define the protected class, although the notion that sex and race are discreet biologically-determined categories is itself a contested one (Butler and Parr, 1999; Woodhams and Corby, 2003). The appeal of these definitions appears to be the simplicity with which members of the protected class are identified. As Woodhams and Corby (2003:165) note, "there may have been doubt about whether the Race Relations Act covered Jews, or whether transsexuals were covered by the Sex Discrimination Act but there was rarely any need for debate as to whether a person was a Jew or a transsexual." Disability, on the other hand, is much more organic and complex, is often hidden, misunderstood and can change over time.

Although many have criticised the definition of disability envisioned in the DDA, it remains to be seen what its effects will be in practical terms. The next section examines this. My interviews with key actors from various service providers, some of whom are disabled and some who are not, reveal the complexities of putting Part 3 into place, and the challenges of ending discrimination in relation to the provision of goods, services and facilities.

(3.3) How Part 3 is being Interpreted and Implemented

In 2002 I interviewed a range of key actors in England, including representatives from private-sector companies, non-profit organisations, government departments, and local authorities. I spoke to the Access Officer for each organisation, although their titles varied from Manager of Facilities and Access to Director of Brand Environment (I refer to them as key actors). In cases where more than one person worked with access issues, I interviewed all of the relevant members of staff. Regardless of their titles, each person I spoke with had primary responsibility for spreading knowledge of the DDA in their organisation. The analysis here is divided into three inter-related themes, (3.4) How service providers understand disability and the DDA, (3.5) How key actors understand and interpret the DDA, and (3.6) What service providers are actually doing, i.e. what policies, practices and procedures are being put into place?

(3.4) How Service Providers Understand Disability and the DDA

Some service providers' perceptions of disabled people can be based on stereotypical portrayals of disability in popular culture – be it in the media, images propagated by patronising charities, or by the government, for example, with its use of the language of 'incapacity' benefit (Adams, 2002). For example, one interviewee said:

> I think the difficulty with the DDA is that people think it's very simple, very straightforward; just bung in a ramp and you've resolved it. Put it into Braille

and it's done. And I think that's the problem really (Sarah, a local authority Access Officer).

This is echoed in the literature by disabled people themselves who feel that these stereotypes are outdated (Gooding, 1994; Morris, 1991; Oliver, 1990). Not surprisingly, these perceptions of disability feed into how service providers tend to think they can make their services accessible. In thinking about "who are disabled people?" most service providers tend to conceive of disabled people as wheelchair users and subsequently equate access with ramps and lifts. Physical access to the built environment is only one form of access, and ignoring the social elements of disablement means that much discrimination will go unchecked, even when disabled people are able to get into the building (Gleeson, 1999).

For example, Sarah, who is her organisation's Access Officer and is disabled, but not a wheelchair user, gave an example of a conversation she had with a reservations clerk at a hotel. As she recounted, "When I telephone for a hotel, and I say I have a disability, they don't let me finish. The next step is 'well we have ramps'." She said she found it irritating and yet felt she had to be patient and grateful, since many hotels she called told her they didn't have any access at all. She persisted and said, "well, that's very commendable, however, my needs are these. I don't like flashing lights in your bathrooms because they cause me to have seizures..I need certain light levels that I can control." The service provider, in this case a hotel clerk, automatically assumed that because the customer was disabled, she needed wheelchair access. Another person noted:

> People can tend to get blinkered in my view just looking at lifts, car park lifts, ramps, revolving doors..Well they get it wrong, undoubtedly, given that a lot of designers and architects like to have spiral staircases in buildings and they love to have all this bright reflective glass everywhere and have lots and lots of steps to make the building look really impressive. You're left with architects and designers who think that's wonderful but have no *direct experience* of the obstacles that they are creating in their wonderful designs for other people who will have problems in the environment. (Emphasis added.) (Margaret, Policy Officer for a large government department)

Given service providers' limited exposure to disabled people and their needs and realities, what do they think of disability vis-à-vis the DDA? Many service providers say the DDA has not penetrated to the core of business interests. One person said the Act is "very hard to understand" and that it is "very, very difficult for businesses to really understand what it means." He expressed that he "had a lot of difficulty getting senior people to understand the commercial implications of the Act."

Because of service providers' general lack of knowledge about disability, the government and the DRC have aimed to provide advice which is straightforward and can be easily understood and implemented. For example, there is a Code of

Practice, which provides advice but does not impose legal obligations, and is not an authoritative statement of the law. However, the Code can be used as part of evidence in legal proceedings, and courts must take into account any part of the Code that appears relevant to questions arising in proceedings. If service providers follow the guidance made in the Code, it may help to avoid adverse judgements in court (DRC, 2002b).

While this has succeeded in getting service providers to think about creating access, it is limited to business terms of reference. The advice in the Code of Practice employs the language of convenience, cost, efficiency and reasonableness. There is a lack of language about empowerment, humanity, kindness, and compassion, which might be expected when talking about interactions between people. Treating people in ways that are morally correct, kind or pleasant, is beyond the scope of liberal law (Razack, 1998). Therefore, creating access is seen as being making businesses profitable along with ending inequality. For example, in *The DDA 1995: An Introduction for Small and Medium-Sized Businesses*, the government says,

> You should keep the duty to make reasonable adjustments under review. It might be appropriate for you to do this whenever you review the efficiency and cost efficiency of your business and working practices (DfEE, 1999:6).

Service providers can also justify refusing service to a disabled person if they can show that they must do so to protect the fundamental nature of their business or service (HMSO, 1995). As a result of this business mindset, some key actors are finding it difficult to persuade their organisations that inclusion is important, not just to meet the minimum requirements of the DDA, or because it's good for public relations, but because equal access should be at the heart of their organisation's interests.

Some service providers are beginning to see disabled people as an attractive group of clientele. In the Code of Practice, and other documents related to Part 3, the DRC (2002b:2) continually repeats that "there are over 8.5 million disabled people in the UK and they have considerable spending power." This message seems to be penetrating through to service providers. As one person said,

> I think if you're making a business case argument, the best thing to do is use the fact that if you don't, you'll lose business, you'll lose money. Given that the spending power of disabled people in the UK is £40 billion, and we have a growing aging population, so more and more people are acquiring impairments. Only about 3 per cent are born with them. Most people acquire them during their working lives or working years. It makes sense if you're running a business to not alienate a large group of your customer base. So it does come down to money (Margaret, Policy Officer for a large government department).

This comment was echoed by others, including Michael, a property manager, who said, of other service providers, "Marks and Spencer and Debenhams have realised that disabled people represent a huge market." Andrew, who manages a telecommunications firm, noted that disabled people are attractive clients because they can tap into "funds and how to access them – disabled people know where that money is." Another person, Spencer, who works for a chain of retail shops, said he's trying to raise awareness within his organisation of "the value it's going to create: 8 million disabled people as potential customers." Some would argue that this prioritising of business interests, and convincing service providers to be accessible so as to generate more business is in keeping with the government's ultimate goal – to keep businesses profitable. David, who works for a large public transport organisation, commented, "as the DDA was written by Tories, it aimed to minimise the burdens on private industry, the Conservatives' biggest supporters.. the government was put into office by businesses so it was nervous about any new costs imposed on business."

This is reflected in the literature. In fact, the 1995 *White Paper* noted that a prohibition of 'indirect' discrimination could have consequences which would be burdensome for businesses (Doyle, 2000). This clearly demonstrates the government's underlying wish to keep the 'burden' on business to a minimum. This seems to be in keeping with some views in the business community, such as this view: "the initial reaction to this legislation will no doubt be that it is an extra headache for businesses" (Field Fisher Waterhouse, 2003:23). Regardless of whether they are doing it to create equality, or because it makes good business sense, there is some indication that service providers are taking on board the duties set out in Part 3. As seen in Section 3.6, the service providers I interviewed are actually taking on many significant access initiatives.

(3.5) How Key Actors Understand and Interpret Part 3 of the DDA

There is frustration and confusion about definitional issues in Part 3. This section allows those voices to be heard, especially the frustrations of key actors from service providers grappling with the Act and trying to change their organisations. However, by the end of this section it will be evident that most interviewees are not allowing their organisations to get mired in these definitional issues. The purpose of this section is simply to illustrate what service providers have to say about Part 3 in their own words.

Key actors feel that there is much confusion about the scope of the DDA, and the definition of disability and 'reasonable adjustments.' Carol, a policy officer from a large government commission, said, "the law is really quite bizarre, since (a) you have to prove you're disabled, and (b) it is not clear what is reasonable and what is not." This does not seem to portend well for disabled people, and in fact, according to a report entitled *Monitoring the DDA*, many applicants and respondents did not understand how the Act defined 'disability' and some disabled people were unsure as to whether or not the Act applied to them (Meager et al., 2002). The definition of disability in the Act has been contentious and 26 per cent of all unsuccessful cases were rejected because the tribunal determined that the applicant was not disabled (DRC, 2003). Meager et al., (2002) were concerned about the burden of proof under the Act, i.e. the fact that disabled applicants have to prove that discrimination has occurred. Very few Part 3 cases went to the county courts, but in terms of definitional issues they seem to be following precedents from Part 2 employment tribunal cases. If so, many disabled people will be excluded simply because they cannot prove they are disabled or provide evidence of the discrimination they faced (DRC, 2003).

One facilities and access manager, George, a buildings control officer, said that the definition of disability, as provided in the Act, is "very wide" and that his organisation "can't cover everyone." This echoes the earlier argument about government trying to limit the burden on businesses. While the definition may be seen as narrow to disabled people, to some businesses it appears too wide, as they feel it will cost too much to implement the DDA.

Not all service providers are so focused on definitions. One key actor noted that she is trying to move her organisation away from the need to prove that clients are disabled. She said,

> I wanted to clearly focus on the can-do approach to invoke the spirit of the DDA. Not just make people prove they have a need before we will provide what they want (Margaret, Policy Officer, large government department).

In addition to the DDA's definition of disability, 'reasonable' is a term which key actors find perplexing. Many pointed out the complexity of users' needs and possible responses. Louise, a policy officer with the government department responsible for drafting the legislation, said, "It's difficult for service providers to get their heads around it..I'm not sure there is an easy answer." Others commented on what they saw as the ambiguity of 'reasonable adjustments.' One key actor said:

> Reasonableness is one of the major problems with it and also to have something that is a law that says it's reasonable to discriminate, to me, is what makes it weak. It either is, or it's not. It is or isn't, isn't it? And that is totally where I think the faults are really. You need to be literally clear with people. This is unacceptable, and this is acceptable. And not have those grey areas (Sarah, local authority Access Officer).

Given the variety of impairments and needs of different disabled people, and the variety of service providers, most interviewees felt that it would be difficult to clear up the ambiguities in Part 3. As Spencer, of the large retail chain, said, the Act "is not very palatable in terms of understanding and implementation." Many of the people I interviewed said the Act was both too ambiguous and too prescriptive. While this may seem contradictory, it indicates the complexity of grappling with what the Act means, and what is required of service providers. As one person said,

> The Act is quite prescriptive. It's not rocket science.... It's grey. It means you sit down with that person and understand the need. Let's forget the Act; let's provide the individual what the individual needs. The Act is just a framework (David, Policy Officer, large public transport organisation).

This complexity, gleaned from several interviews, indicates that the Act is misunderstood and misinterpreted, possibly because it introduces relatively new ways of thinking about disability discrimination, and because of the inaccessibility of legal language itself. As Spencer, who works for a large retail chain, said, "the Act is pure legalese." At the end of the day, key actors say that creating accessibility entails good customer service. As one person said,

> What we want to do in (organisation name), is to ensure that our services are accessible, in the best possible way, and not worry too much about ramps and lifts and those sorts of things. At the end of the day it's about what the customer wants, what they need, when they need it. Those are the needs that we try and meet (Margaret, Policy Officer, large government department).

This approach is reflected by most key actors. While they find the language and definitions of the DDA confusing, they are committed (some of them passionately) to making their services more accessible, and to spreading a spirit of inclusion within their organisations. To understand how they are undertaking this task, the next section considers what policies and practices service providers are implementing in response to Part 3 of the DDA.

(3.6) What Service Providers are Doing

Service providers in Nottingham, Manchester and all over London provided examples – in interviews as well as documentation – of actions being taken to increase disabled peoples' access to goods and services. These ranged from changing policies to making improvements to physical access to buildings, providing services in alternative ways and formats, increasing staff awareness through training, and creating new technologies to serve disabled customers.

In addition to interviewing key actors, I corroborated the claims they made in interviews by obtaining documentary evidence of the policies and practices they were creating, seeing physical accessibility changes myself, triangulating information given to me in interviews with other institutional actors and obtaining statistical data on the numbers and proportions of disabled customers and employees, for example, and the amount of capital to be spent on DDA-related changes and specifically what these funds were earmarked for. Therefore any claims made by key actors in this section have been verified by additional sources.

One local authority, for example, has adopted a social model of disability which goes beyond the DDA's medical model. This is stated by one of their Councillors, who is the Lead Member for Equalities and Inclusion in their publication 'DDA 1995: Our Responsibilities (Handbook).' While this may seem philosophical, the key actor, who is the local authority's Access Officer, assured me that the borough is committed to inclusion in the broadest sense – not just what the DDA requires, and not just in terms of disability but also in terms of gender, race and age. This local authority represents the poorest borough in London per capita, but conversely is culturally very diverse, with a sizeable Bangladeshi population. The key actor emphasised the fact that all of her training programs on disability, for example, include sensitivity to and awareness of the particular cultural contexts of Bangladeshi Britons, and how these affect their experiences of disability. This is an example of practice which goes beyond liberal abstract notions of individuality, where differences – be they based on culture or ability – are incorporated into policy in meaningful and substantive ways.

Another organisation which is taking 'access' on board is a large public transport organisation which is responsible for providing and regulating underground trains, buses and taxis for Londoners. They have published 'The Mayor's Transport Strategy: Executive Summary and Accessibility Action Plan,' which goes beyond simply complying with the DDA, and is linked to one of the wider goals of the Greater London Authority (GLA), namely that London should be a 'fair' city (Greater London Authority, 2002). The Plan talks in depth about ending social and attitudinal barriers, and says that services provided to disabled customers (or elderly customers or those carrying children) should be convenient and comfortable, far more palatable than the DDA's talk of ensuring services are not 'unreasonably difficult' to access. The Mayor's Plan uses terminology such as "empowerment, inclusion and decision-making," again a far cry from the DDA's suggestion that service providers merely 'consider' the needs of disabled customers. This organisation has also made all new buses accessible (ahead of the regulatory deadline in Part 5), is phasing out all old 'Routemaster' buses, and offers a door-to-door transportation service for customers who need it, blue badge parking, and other personalised services.[6] Most importantly, they have exhibited, in all of their literature, openness to the suggestions and needs of disabled customers.

A much more 'high-tech' service, but one which will nonetheless help some disabled customers, is the introduction of new text phones offered by a large telecommunications company. This company has created 2000 of these new machines, whose software will give text payphone customers sending typed or voice messages immediate communication either with 'textphones' – terminals used by deaf and speech-impaired people – or standard phones. The service is called Text Direct, and it automatically introduces a Typetalk operator into the call when a textphone user is connected to a voice user. Typetalk is a national service funded by BT and is run by the Royal National Institute for Deaf People (RNID). An operator reads a caller's typed message to a hearing person and then types the

---

6   Following his election as Mayor of London in 2008, Boris Johnson pledged to reintroduce Routemaster buses despite concerns about their safety.

reply to the caller. The company says there are 150,000 profoundly deaf people in the UK, and that the addition of Text Direct to their text payphones underlines their commitment to meet the needs of all of their customers.

In terms of the built environment, many large service providers are working to make their buildings accessible to everyone, including ambulant disabled people and people with sensory and mental impairments. A large chain of retail shops said that almost one hundred percent of their shops were fully physically accessible, and the remaining few would be accessible in advance of October 2004. One commercial property management company whose clients include the government departments in Whitehall, have incorporated access improvements to buildings into their ongoing systematic maintenance upgrades. In this sense, access has become a part of 'business as usual' for them; they have followed the Code of Practice's advice to treat access issues as central to their business plans.

Finally, a large government department has taken access issues on board from the highest level. In addition to making all of their buildings physically accessible where possible, they have created a campaign to win "hearts and minds," as their key actor articulated. The department has extensive training programmes, not only for new staff, but also ongoing education on inclusion for all members of staff. These go beyond disability awareness and good customer services practices to include two key aspects: (1) how individual staff members can personally ensure inclusion of disabled people, and (2) how teams and units can work together to create inclusive environments for disabled people in their workplaces. Additionally, the department has linked this training into its broader work to end discrimination – including equality in terms of gender and race. This broader work is about recruiting more staff of diverse backgrounds, be they disabled, women or minorities, not just into the department, but also into progressively higher management positions. The key actor says that her remit is to create inclusion, and as such, goes far beyond just meeting the requirements of the DDA.

These are just some examples of what service providers are doing – in terms of new policies, practices and procedures – to be inclusive of disabled people. This seems to be different from simply ending discrimination. Indeed, key actors from all of the above organisations were emphatic that their work on inclusion is not limited to the DDA. While the presence of such a law helps strengthen their case for changing their organisations, all of them say that the substance of the DDA, in terms of what is considered reasonable or not, had no bearing on their work. One person even went so far as to say that even without the DDA, his company would have done the same work on inclusion because "it's the right thing to do." Key actors argued that perhaps the only way in which the DDA served as a useful tool was as a 'stick' to remind their organisations of the threat of litigation should they not be inclusive. Most key actors rarely had to use this stick because of a much more palatable idea: the one that being more inclusive is simply the right thing to do and makes good business sense. Not surprisingly, key actors are passionate about creating inclusion in their organisation, and though they have faced and continue to face resistance along the path to inclusion, they have used such moments of resistance to educate colleagues about disability issues. In some sense then, the DDA has been of less use to service providers, substantively speaking, than the mere awareness of its existence.

(3.7) Conclusions

This chapter provided specific evidence about how some service providers and their respective key actors view disability and the DDA, and about what they are doing in response to it. For example, according to a recent Department of Work and Pensions (DWP) report, 65 per cent of respondents were unaware of the 2004 Part 3 duties (DWP, 2002). This differs from my findings, in which all twelve of the key actors I interviewed were clearly aware of their organisations' 2004 duties under Part 3. Perhaps this can be attributed to the fact they are leaders among service providers, and of course, because of the small sample size of the study, they cannot be said to be representative of all service providers.

What can be said about the attitudes encountered during interviews is that they mirrored some of the wider public attitudes about disability, if only because they are based on popular perceptions and stereotypes (Morris, 1991). There is a need to go beyond simply 'considering' what the needs of disabled people might be (as the *Introduction for Small and Medium-sized Businesses* suggests) or even merely consulting disabled people regarding their access needs. What would be more egalitarian and emancipatory would be to have disabled people in positions of power, and not just as key actors, but working at management levels within large businesses, government departments and local authorities. While the employment of disabled people is within the remit of Part 2 of the DDA, there is no evidence that disabled people are attaining management positions in service providing organisations on a significant scale. On the contrary, disabled people are still disproportionately under-represented in employment generally, let alone in terms of management positions (Woodhams and Corby, 2003).

My evidence is perplexing; on the one hand there seems to be uncertainty and confusion among service providers about the definition of disability under the DDA, and about what are considered 'reasonable adjustments.' At the same time, the examples cited in the previous section show that service providers are making their services more accessible, and in some cases, going beyond what they believe is required of them under Part 3. This illustrates a paradox about the DDA – service providers are unclear about what it really means or entails but they are making adjustments to their operations, regardless of this uncertainty. Some argue that the DDA is simply too limited in its substance and scope to have any significant positive impact on disabled people (Oliver & Barnes, 1998). Others argue that the DDA is an adequate tool for creating inclusion, because the scope of law itself is limited and ADL can only go so far in ending inequality. As one interviewee said:

> The DDA is a reasonable attempt, it's about as far as you can go, legislatively.. discrimination is an attitude of the mind. You can remove physical barriers; you can't remove attitudes (Graham, Policy Officer, government department).

In so much as it works to ensure that goods and services already available to able-bodied people be made accessible to disabled people, Part 3 is having an impact. For a more radical goal of fundamentally changing the nature of service provision to make it inclusive in its inception, rather than simply providing access to a pre-existing set of goods and services, it is not certain Part 3 can or will do this. In fact, the law is so limited that the government announced in January 2003, at the start of the European Year for Disabled People, that it would introduce new legislation in parliament to supplement and strengthen the DDA (RADAR, 2003). Secretary of State Andrew Smith said the Bill "would significantly advance the rights and opportunities of disabled people up and down the country" and would include measures proposed by the Disability Rights Task Force (RADAR, 2003:3). Some of these include: extending the definition of disability to include people with progressive illnesses such as cancer, HIV and multiple sclerosis, removal of exemptions from small employers and certain categories of employment, and a placing a legal duty on public authorities to promote equality of opportunity for disabled people (RADAR, 2003). Currently, there is no such duty to promote equality, only one to make 'reasonable adjustments.'[7]

Because of the uncertainty about what is considered 'reasonable,' the outcomes, in terms of access for disabled people, are uneven, and different service providers are doing different things based on their particular knowledge, resources, and understanding of disability and the DDA. For example, while the transportation organisation may be doing exemplary work, one cannot be sure that transportation authorities in other regions of the UK are also taking the DDA on board. That is outside the scope of this study, but it is important to consider whether a disabled person can expect any linking up of individual services, or consistency in service provision they receive from one place to the next.

For example, Part 3 only addresses individual service providers in their particular locations and buildings, but not the built environment as a whole. As for what motivates service providers to increase their accessibility, it seems that the exemplary cases I mentioned in the previous section are doing this either because it fits into the broader mandates or cultures of their organisations, or because of the 'goodwill' or public relations-savvy of particular individuals, or because disabled people within the organisation, such as key actors, have fought long and hard for such changes. The results for disabled people are uneven from organisation to organisation and difficult to quantify. As such, Part 3 of the DDA has caused service providers to focus on how their organisations can become more accessible, but this is isolated from wider debates among disabled people or the general

---

7 This has now been partially addressed by the Equality Act [2010]. For further details, please see the Epilogue.

public about issues of access and equality. Larger issues of structural inequality, or systemic bias or exclusion, for example with regard to the built environment, are not considered, discussed, or acted upon (Doyle, 2000; Gooding, 2000).

Part 3 of the DDA does not fundamentally alter societal inequalities against disabled people, but rather attempts to make it easier for disabled people to access the socio-economic structures which are already in place and have been built by able-bodied people without disabled people in mind (Oliver and Barnes, 1998). My empirical evidence indicates that Part 3 is limited because it does not envisage a new world in which all members of society are equal participants, but rather, it seeks to maintain the current society but in a more accessible fashion, and only in a manner which service providers consider 'reasonable.' This highlights one of the key themes of this book, i.e. that the concept of 'access' is itself a limited one, and must be expanded and broadened, if inclusion is the true goal of civil rights legislation.

This chapter illustrated two main points. The first was that law is context-dependent, and because of the different ways in which it is being interpreted and implemented, Part 3 will result in very different outcomes for disabled people depending on their particular geographical contexts. The second theme shown is the limited scope of Part 3 due to its liberal underpinnings. This legislation does not aim to change people's perceptions of disability or their attitudes towards disabled people. For example, the Minister for Disabled People acknowledged, in an address, that "legal requirements alone won't change it, we have to win hearts and minds" (Eagle, 2002). Systemic inequality and deeply ingrained attitudes require much more than Part 3 to be uprooted. As Barnes and Oliver (1990:114) note,

> Even fully comprehensive and enforceable civil rights legislation will not, by itself, solve the problem of discrimination against disabled people. This is because, like racism, sexism and other forms of institutional prejudice, discrimination against disabled people is institutionalised in the very fabric of British society. It encompasses direct, indirect and passive discrimination. It has its roots in the very foundations of western culture.

How can institutional prejudice be overcome? What is institutional discrimination and how do disabled people experience it? How can institutions transform themselves so as to be rid of systemic prejudices? As Doyle (1996:299) notes, antidiscrimination legislation in Britain, which is centred on individual-based complaint processes, such as the SDA 1975, RRA 1976, and DDA 1995, is considered inadequate for eliminating "systematic and institutionalised discrimination." Systemic discrimination is a much more complex phenomenon than individual discrimination, and the next three chapters will examine issues of institutional discrimination in the higher education sector.

PART II
Implementation

Chapter 4

The DDA and the New Managerialism in Higher Education

(4.0) Introduction

Since the 1980s there has been a shift in the operation of UK public institutions from a welfare-based model to a more individualistic, consumer-driven 'market' model (Henkel & Little, 1999; Kogan & Hanney, 2000; Newman, 2001). This has occurred in relation to all aspects of public policy, including health, education and welfare (Newman, 2001). In this new mode of public management, which Clarke and Newman (1997) call 'managerialism', free market paradigms and mechanisms have been introduced into the public sector as a result of government policies. Such changes have not been limited to the UK. For example, reductions in government welfare functions, coupled with the increased use of language and practices from 'management' in public sector institutions, have also occurred in Australia, the USA, and elsewhere and indeed seem to recur cyclically in times of economic retrenchment (Zifcak, 1994).

This managerial shift has not been accidental, or simply dictated by market forces, but it reflects new philosophies in which governments have become increasingly committed to an agenda to make public institutions more efficient, accountable, and even profitable (Zifcak, 1994). Moreover, this shift is not just reflective of concerns about the costs of running the public sector, but indicates a change in philosophies about the role of political and legal systems in shaping society and individuals. This has included debates about what constitutes society, about the value of free education and free health care, and debates about the role of government in peoples' lives.[1]

Managerialism has created significant challenges for higher education institutions (HEIs, which include universities, higher education colleges and further education colleges for the purposes of the DDA). As Kogan and Hanney (2000:11) note, "perhaps no area of public policy has been subjected to such radical changes over the last 20 years as higher education." HEIs are experiencing ever-

---

1 For example, former British Prime Minister Margaret Thatcher once described her ideals as, "compassion and concern for the individual and his freedom; opposition to excessive state power; the right of the enterprising, the hardworking and the thrifty to succeed and reap the rewards of success ..." (in Cosgrave, 1978:68). Evans (2004:3) also notes that Thatcher was against "state interference with individual freedom; state initiatives that encourage an ethos of 'dependency'; woolly consensuality ...".

increasing funding challenges, and have to comply with increasingly centralised, bureaucratised and highly complex legal requirements such mandatory educational oversight by the Quality Assurance Agency (QAA) for Higher Education, the Higher Education Funding Council for England's (HEFCE) Research Excellence Framework (REF) exercise, and mandatory reporting to Office for Fair Access (OFFA).[2] This means that institutions have to do more with less, and they have to work harder to obtain financial resources in an increasingly competitive market-like sector (Kogan and Hanney, 2000).

In this context, the introduction of Part 4 of the DDA in 2001, also known as the Special Education Needs and Disability Act (SENDA), is another legislative requirement placed on HEIs. I argue – using findings from a postal survey of disability officers – that this legislation is in keeping with the spirit of new managerialism in the public sector. I divide the chapter as follows: in (4.1) I highlight some of the key provisions of Part 4 of the DDA, in (4.2) I outline new managerialism, in (4.3) I consider Foucault's theories of power, and in (4.4) I examine managerialism in the context of (a) Audit cultures and accounting imperatives, and (b) The managerialisation of local actors. In (4.5) I conclude the chapter.

(4.1) The DDA's Liberal Discourses

As noted in Chapter 2, the DDA and its constituent parts (Parts 2, 3, 4 and 5) are based on liberal notions of individual discrimination, and medical definitions of disability. In this section I will briefly examine the legal provisions of Part 4 of the DDA, hereafter referred to as SENDA, to highlight what the higher education sector must do to meet the requirements of the law, and to help examine, later in the chapter, whether SENDA will provide a route to inclusive education. Inclusive education is defined as maximising the participation of all students, and removing environmental, structural and attitudinal barriers to their participation (UNESCO, 2000). In order to be inclusive according to this definition, SENDA would have to force HEIs to meet the needs and aspirations of all students, regardless of ability, by changing physical and social environments.

Before detailing SENDA, it is important to note the complex and sometimes overlapping nature of the DDA's various sections. For example, HEIs employ a significant number of staff members and therefore the employment provisions of the DDA Part 2 also apply to them. Most institutions also provide non-educational services to the public, such as conference facilities, accommodation, catering and consultancy services. Because these are considered goods and services, HEIs are accountable under the requirements of Part 3 in relation to these services. This means that HEIs have to be aware of Parts 2, 3 and 4 of the DDA, although in

---

2 For example, since September 1998, university students' tuition fees are no longer fully paid for by the government, and top-up fees will be in effect from September 2006. For a discussion of top-up fees, introduced in 2004, see Taylor, 2005.

many institutions there is no overall strategy for access, and different departments are responsible for complying with different parts of the Act. For example, one disability officer noted that she receives enquiries from members of staff about Part 2 provisions of the DDA, and that even though this is not officially part of her role, no one within the university has clear responsibility for this.

In the DDA legislation created in 1995 (Parts 2 and 3), the only statutory obligation for higher education was for funding councils to have regard for the requirements of disabled people in exercising their functions, and to make the publication of a disability statement by each HEI one of the conditions for receiving grants (HMSO, 1995). This had little measurable impact in the higher education sector, partly because the phrase 'to have regard for' is considered ambiguous and unclear (Palfreyman and Warner, 2002). Additionally, the Higher Education Funding Council for England (HEFCE) had done little to enforce, analyse or utilise the disability statements provided by HEIs (Adams, 2002). In fact, there is no punitive mechanism in place for HEIs that do not submit such statements. As a member of staff from HEFCE's National Disability Team noted in interview: "I would like to see HEFCE suspend payment if they didn't receive a disability statement. It would never happen."

Therefore, the DDA 1995's minor provision for higher education proved to have little effect on the sector. Because the Act was seen to be lacking in its consideration of education, the government's position changed after the Disability Rights Task Force published its report *From Exclusion to Inclusion,* which recommended the inclusion of further and higher education within the DDA's legislative framework (DRTF, 1999). In May 2001, the government introduced SENDA, for which the basic duty to provide reasonable adjustments came into force in September 2002, the duty to provide auxiliary aids and services in September 2003, and the duty to make adjustments to physical features in September 2005. The corresponding implementation dates for Part 3 of the DDA were 1996, 1999, and 2004. This is relevant because HEIs have duties under Part 3 for some of their services which are used by members of the public such as conference and dining facilities. As with Part 3, SENDA's staggered dates of implementation were designed to give institutions time to devise strategies and find resources for increasing accessibility.

How does SENDA work? As of 1 September 2002, SENDA states that it is unlawful for the governing body of an HEI to discriminate against a disabled person as defined in the Act, in the following ways: (1) in the arrangements it makes for determining admissions to the HEI, (2) in the terms of admission, or, (3) by refusing or deliberately omitting to accept an application for admission (HMSO, 2001). It is also illegal to discriminate against a disabled student in the non-academic student services it provides or offers to provide, or to discriminate against a disabled student by excluding him or her permanently or temporarily from the HEI. 'Student services' are any services that are provided wholly or mainly for students apart from the provision of education itself, including accommodation, catering and leisure facilities.

The definition of discrimination under SENDA is similar to the one found in Part 2 of the DDA, although there are differences in relation to the concept of 'reasonable adjustments.' Under SENDA, the governing body of an HEI is said to be discriminating against a disabled person if, for a reason relating to that person's 'disability', the HEI treats the disabled person less favourably than it treats, or would treat, others to whom that reason does not or would not apply, unless the treatment in question is justified.[3] As with Part 3, this defence of justification has been criticised for weakening the rights of disabled people, as HEIs can "hide behind 'academic standards'" and the law "permits discrimination in various circumstances" (survey respondents, 2002).

One interesting difference between SENDA and Part 2 is that the duty to make reasonable adjustments is a duty to disabled people at large rather than in relation to specific disabled persons. This differs from the DDA's employment provisions and means that HEIs need to consider the provisions they make for disabled people on a wider scale than simply the individual level. Initial indications were that there is widespread anxiety about this duty in the sector (Lewis, 2002). Some analysts say that universities are taking risk management approaches, weighing their options and hoping to do as little as possible until case law determines how accessible they must become (Adams, 2002).

A final section of the law which has garnered much attention has to do with the disclosure of a student's impairment. According to SENDA's disclosure provision, if one person at an HEI knows about a particular student being disabled, then that institution is deemed to know. This makes institutions uneasy because it means that those who are informed of students' impairments will have to convey that information to others in the institution, which could potentially breach confidentiality. According to several survey respondents, this appears to contradict the Data Protection Act of 1998, which places clear demands upon those holding personal data in terms of the security that must be applied to protect it (HMSO, 1998). As one survey respondent noted, "problems arise with providing resources/ appropriate staffing for very late applicants and for students who do not wholly disclose they have a disability." Issues around confidentiality and disclosure are also indicative of the stigma attached to disability, especially for people with intellectual impairments. Many students do not disclose their impairments because they are concerned that identifying as being disabled will negatively impact their academic and/or career opportunities (Holloway, 2001).

Like other parts of the DDA, SENDA does not fundamentally call into question power relations between disabled and able-bodied persons. For example, there are no requirements to reconfigure existing institutional governance structures which are outmoded, exclusionary and generally the domain of able-bodied people (Barnett, 2002; Morley, 2004). In section (4.4) I evaluate two main ways in which

---

3   The medicalised definition of disability is discussed later in the chapter. According to the social model, a disabled person is described as having impairment, not disability (Gooding, 1994).

the new managerialism in HEIs impacts upon the implementation of SENDA. First, the law itself has liberal/new managerial foundations, such as its reliance upon direct discrimination[4] (rather than indirect or systemic discrimination,[5] or oppression[6]), and the recommendation that HEIs undergo audits of their institutions.[7] Second, the responses to the legislation from HEIs are mitigated by these institutions' modes of operation which are increasingly new managerialist in scope. I will now examine new managerialism before detailing the two themes set out in the introduction.

(4.2) What is New Managerialism?

The emergence of a managerial reform movement in the public sector, also referred to as the New Public Management (NPM), has been evident in Britain since the late 1970s (Clarke, et al., 2000). British governments of both major parties – the Conservatives and Labour – have pressed ahead with these managerial reforms in an attempt to better manage the administrative machinery of government, and to appease a critical public who appear to be losing faith in the bureaucratic welfare state (Clarke and Newman, 1997; Zifcak, 1994). For New Labour, as the party was re-branded under Tony Blair, modernisation was a key element of NPM. During the period of the New Labour government, from 1997 to 2010, Britain's public services, including the NHS, local government, the criminal justice system, and education, underwent significant programmes of modernisation, followed with a similar approach by the Conservative-Liberal Democrat coalition government elected in May 2010. Indeed, modernisation has become the mantra of the government's ongoing efforts to reform public services. As Tony Blair said, "if government is going to be effective at delivering services in the way people want them for today, it has to be modernised" (in Newman, 2001:83).

David Cameron, the current Prime Minister, also emphasises the importance of modernising public services when he told the Welsh Assembly: "I believe now is the time to modernise our public services – and in England, that is what we're doing."[8] Modernisation is seen as unquestionably positive – the way forward to

---

4   Direct discrimination happens when disabled people are treated less favourably than able-bodied people are treated or would be treated (HMSO, 1995; HREOC, 2003).

5   Indirect or systemic discrimination happens where a rule or situation unreasonably excludes or disadvantages disabled people in practice, even if the intent is not to discriminate (HREOC, 2003).

6   Oppression is much more pervasive and yet subtler than direct discrimination (Young, 1990). For example, social norms of disability and cultural portrayals of disabled people can be oppressive but not necessarily quantifiable or measurable as forms of direct discrimination.

7   SENDA's Code of Practice recommends that HEIs undergo institutional audits (HMSO, 2001).

8   bbc.co.uk, 12 July 2011.

"revolutionise public services in England and improve lives."[9] What lies behind the term modernisation? For Cameron, modernisation is about choice, a free market approach to the provision of public services such as opening up schools and hospitals to private sector competition, for example. Modern is understood to mean: *the way things are done in the private sector.*

Modernisation is presented as a rational, forward-looking effort to renew public services to meet the new demands of the public and the needs of the business world and the free market. However, many writers argue that far from being value-neutral and non-ideological, modernisation, as exemplified by government programmes, embodies specific values, assumptions, and worldviews (Clarke and Newman, 1997; Monbiot, 2000; Newman, 2001; Zifcak, 1994). Modernisation, in terms of British public policy, is embodied in neo-liberal reforms aimed at opening up the public sector to market mechanisms and positioning members of the public as consumers (Newman, 2001). As Newman (2001:83) says, "there was a continued focus on organisational efficiency and performance, and on the search for business solutions to social and public policy problems."

Efficiency is viewed by governments as a neutral and technical matter, rather than being tied to particular interests (ibid.). For example, a key New Labour initiative was the modernising of public services, a goal which ministers believe can be achieved through efficiency measures. As noted in a report of the government's *Comprehensive Spending Review*:

> The Government is determined to improve the quality of services. That will require greater efficiency to get more out of the money which is spent on services (Archives, 2005:1).

Indeed, in 2003 the Chancellor and Prime Minister commissioned an efficiency review of all UK government expenditure and a Spending Review was also conducted in 2010.[10] Such measures are referred to as if they have no political implications, and as if all considerations of how to achieve efficiency are free of particular claims or value judgements. However, this view is contested. In his book, *Foucault and Education: Discipline and Knowledge*, Ball (1990:154) notes, "efficiency itself is taken as a self-evidently good thing." From a Foucauldian perspective, management is a professional and 'professionalising' discourse which allows those who use it to make claims to management expertise, organisational leadership and decision-making, and to create procedures which cast 'others' as objects of that discourse and as recipients of those procedures (ibid.).

Foucault (1980) argues that like other professional discourses, management produces the object about which it speaks, namely the *organisation* (in Ball, 1990).

---

9   Ibid.

10   As of April 2005, all Government Departments are required to publish Efficiency Technical Notes (ETNs) about how they are going to measure efficiencies and achieve the targets set for them. For a specific example, see Prentis, 2005.

He says that the discourse of management embodies a clear empiricist-rationalist epistemology in which organisational control and the actions of individuals are contained within a technocratic perspective (ibid.). In this view, social life can be mastered, measured and understood rationally and scientifically. Management discourses present themselves as objective, neutral mechanisms dedicated to efficiency: "the one best method" (Ball, 1990:156). Ball (1990:156) notes that management is an "imperialistic" discourse which views the social world as irrational and chaotic – a world which needs to be brought into the "redeeming order" of management.

How does this ideology manifest in practical terms? Newman (2001) notes that governments have actualised these ideologies of management in two main ways: through the use of market mechanisms, and by increasing the state's regulatory capacity. She notes a shift in the government's role towards regulating services, setting standards, and monitoring quality (ibid.). Newman (2001:84) describes the previous government's managerial approach in the following way:

> Labour has extended this process through its use of performance targets, standards, audits, inspection and quality assurance schemes, all backed by additional powers for government to impose mandatory measures on organisations deemed to be performing poorly.

This highlights a key difference between Labour's approach to managerialism, and that of previous conservative governments. While the governments of Prime Ministers Margaret Thatcher and John Major were guided by market-driven ideologies, and believed in both the sale of public assets to the private sector, and the importing of private sector philosophies into public institutions, New Labour sought to more actively control and centralise public sector institutions and activities. As May et al. (2005:704) note, the new Labour regime "has witnessed the development of ever tighter regulatory controls aimed at securing the self-regulation of non-statutory welfare providers and welfare recipients."

Labour governments from 1997 to 2010 also gained more centralised control of local government, the voluntary sector and educational institutions by introducing a pragmatic philosophy of *Best Value* to replace the older mechanism of *Compulsory Competitive Tendering* (CCT) (May et al., 2005). While the stated goal of such re-organisation may be to decentralise power and create more local control over purchasing decisions, for example, the effect is that

> New Labour have sought to exert a much greater control over how non-statutory "partner" agencies deliver welfare services, with a new tendering process effectively dictating the policies and procedures an agency must follow in order to enter into a service contract (May et al., 2005:709).

In a paradoxical way, New Labour increased local control over day-to-day management of non-state agencies whilst increasing the central government's role in overseeing these local decision-making processes, and enforcing regulation through the use of financial and administrative levers (ibid.). As May et al. (2005:710) note,

> Given the degree to which Best Value determines the regulatory framework, however, it is difficult to read such a move as evidence of genuine de-centralisation. Rather, we appear to be witnessing a re-centralisation and formalisation of state power...

This new managerialism has resulted in significant shifts in the autonomy of organisations in the public sector. While higher education organisations are, strictly speaking, private organisations, because of their reliance on governments as primary sources of funding, and their educational roles, they are seen as part of the public sector. With these changes, public perceptions of the higher education sector are also shifting. Scott (2000) notes that there is increasing debate about whether higher education is a public service or if it has more utility for the private sector by serving as part of the knowledge industry.[11] The impacts of managerialism in higher education are evident (Scott, 2000). For example, the term 'management' has become ensconced in the day-to-day workings of institutions (ibid.).

The desire for good management is a common goal for schools, colleges and universities, and is established as the best way to run educational organisations (Ball, 1990). As Ball (1990:156) notes, this transformation has shifted the governance of HEIs from professional/collegial to managerial/bureaucratic modes of operating, and the higher education sector is "increasingly subject to the logics [sic] of industrial production and market competition." Like other managerial government initiatives, the DDA also places reliance on the market to correct the problem of discrimination against disabled people (ODPM, 2002). The next section will analyse Foucault's theories of power and their usefulness in understanding HEIs.

(4.3) Foucault, Power and Higher Education

> Universities are the most conservative institutions when challenged for change. They mirror old monasteries of the Middle Ages as 'custodians of the truth.' Issues of class structure and power are intertwined with their nature (postal survey respondent).

As evidenced in (4.2), HEIs in England and Wales are undergoing managerialist shifts as part of a wider modernising agenda in the public sector. These shifts have two main effects in relation to notions of power in institutions. The first is a

---

11   Given the trebling of university tuition fees in England and Wales in 2012, there is significant debate about the role of the HE sector in the UK economy.

centralisation of power, with regulatory bodies governing HEIs, and centralising tendencies within HEIs themselves. The second, emanating from the first, is the enforced self-regulation of individual actors and units within HEIs. Foucault's (1979) theories on power can be used to illustrate and illuminate both of these effects. As Hekman (2004:200) notes:

> In the course of his analysis, Foucault identifies several new forms of this power rather than one. This in itself is significant. Foucault is not arguing, as did Marx, that power has shifted from one single location to another. Rather, he argues that power has become diffused. Instead of emanating from a single source, it is spread throughout every corner of society, informing the social structure as a whole.

Foucault traced the genealogy of these new forms of power that he identified as 'pastoral power' to the institutions of Christianity, starting in the Middle Ages (ibid.). He claimed that the church's practices envisaged new forms of power in which the individual became the locus of that power:

> This form of power applies itself to immediate everyday life which categorises the individual, marks him by his own individuality, attaches him to his own identity, imposes a law of truth on him which he must recognise and which others have to recognise in him (in Dreyfus & Rabinow, 1983:212).

Foucault (1983) argued that this new form of 'pastoral power' became integrated into the contemporary western state, so that it became "a modern matrix of individualisation, or a new form of pastoral power" (in Dreyfus and Rabinow, 1983:215). While Foucault has written volumes about pastoral power, his most significant argument, employed in this chapter, is that non-state institutions have been employed to enact pastoral power. As Hekman (2004:200-201) notes: "as the western state takes on a radically new function, overseeing the individuality of its subjects, the locus of power is diffused beyond the state to the institutions of civil society."

In the context of HEIs this allows us to examine what disability officers say about their institutions in relation to Foucault's notions that power is not traceable to one location, but that it is everywhere. Foucault (1979) detailed his theories of power in *Discipline and Punish*, which noted that 'disciplinary power' exemplified a particular form of pastoral power. In this seminal work,[12] he posited that the particular forms of state power materialised in prisons changed how power itself was constituted. Noting the challenges of locating these diffuse new forms of power, Foucault (1979:27) argued that disciplining relations "go right down into

---

12 *Discipline and Punish* (1979) is seminal because its impacts went far beyond the realm of prisons and corporeal discipline about which it is written. For example, see Gane and Johnson, 1993, who discuss the impact of Foucault's theories in various academic disciplines.

the depths of society, that they are not localised in relations between the state and its citizens."

In HEIs these power relations are manifested throughout institutions – in relations between and among lecturers, administrative staff and students – and in complex ways which defy simple categorisation. HEIs are both highly structured, and yet are operated by means, modes and systems which are informal, discreet, and not always transparent, even to actors within them (Ball, 1990). Two main themes emerge from Foucault's arguments on disciplinary power. The first is that disciplinary power is generally incompatible with traditional notions of sovereign power (Foucault, 1980). Disciplinary power is not a relation between a sovereign and its subjects but a network of power relations that permeates society (ibid.). Foucault (1980:39) uses the notion of the capillary to describe this:

> But in thinking of the mechanisms of power, I am thinking rather of its capillary forms of existence, the point where power reaches into the very grain of individuals, touches their bodies and inserts itself into their actions and attitudes, their discourse, learning processes and everyday lives.

The second main Foucauldian theme in relation to disciplinary power is the idea of normalisation. Foucault (1979) argues that distinguishing between what is normal versus abnormal is central to these forms of power. Like the capillary, the "normalisation that is the goal of disciplinary society is everywhere" (Hekman, 2004:201). Power, according to Foucault (1979), is a series of relations that extends throughout society. In analysing responses to my postal survey, I examine not the origin or roots of power, for they are not traceable to a single location, but rather, the effects of power, the 'points' of power that are manifest in social relations (Foucault, 1980).

This is why it is important to examine the micro-social elements of institutional life, so as to gain an understanding not just of institution-wide policies and practices but also of the subtle power relations exercised at the level of the daily, the ordinary and even the mundane. It is these realms that must be unearthed, exposed and interrogated.[13] A final body of work of Foucault's (1980), which aids this analysis, is his notion of 'governmentality'. In this theory, governmentality encompasses multiple spheres, from government, to administrative institutions, all the way down to individuals (Foucault, in Dreyfus and Rabinow, 1983). As with his theories of disciplinary power, governmentality is a mentality and an ethos that pervades all aspects of life (ibid.). For, example, the subject of governmentality is not the legal or 'juridical' individual with legal rights but rather a subject in relation to others, subjected to multiple forms of discipline, both within state apparatuses and beyond (Foucault, in Martin and Gutman, 1988).

---

13   For a detailed discussion of unearthing and interrogating values in higher education see Chapter 5. For a detailed discussion of micro-social elements of institutional life, please see Chapter 6.

Foucault's theories of power are central to my arguments that SENDA can best be understood in the context of how institutions work, and in uncovering power relations and networks within HEIs. It cannot be assumed that SENDA will necessarily or automatically lead to the inclusion of disabled students. Rather, this chapter examines how SENDA is being received and responded to in the context of increasing managerialism. This is done with an understanding of Foucauldian notions of power, which illustrate that power is diffuse, difficult to trace, and a challenging aspect of institutional life. The next section examines managerialism at work in the context of how HEIs are responding to SENDA.

(4.4) Managerialism in Action

In this section of the chapter, using responses from a postal survey of disability officers in England and Wales, I explore the interrelationships between the NPM and the varying responses to SENDA in institutions. I do so in relation to two themes: (a) Audit cultures and accounting imperatives, and (b) The managerialisation of local actors.

Survey Sample

Postal surveys were sent to 296 named disability officers at HEIs in England and Wales and 118 responses were received, providing a 39.8 per cent response rate. The respondents include a broad range of institutions, including Russell Group Universities, Red Brick Universities, New Universities, and Colleges of Further and Higher Education.[14] These institutions are geographically spread across England and Wales, and include those in urban, rural and suburban settings. The respondents held a variety of job titles, including Head/Manager/Director (38), Team Leader (5), Officer/Coordinator (26), Adviser (11), Assistant (3) and 3 people with other academic titles (Dean/Lecturer/Registrar). In total, there were 88 different job titles. Job titles denote not only practical aspects such as pay scale, seniority and benefits, but also have much symbolic value in the higher educational system which has historically been very prestige-driven and title-obsessed (Barnett, 1990).[15]

There was variation in office size, and numbers of staff for disability issues, ranging from offices staffed by one person, to the largest number, 150. The majority of offices (more than 60 respondents) had a staff complement of between 1 and 10 people. Annual budgets of these offices ranged from nought to £2,000,000, with

---

14 For more detailed information on the typologies of institutions, see Chapter 5.

15 For example, Marie Trottier, University Disability Compliance Officer for Harvard University, said in interview that one of the reasons she believes her phone calls are always returned promptly is that she leaves messages saying "Hello this is Marie Trottier calling from the President and Provost's Office." This is an example of prestige having material impacts on disability officers' efficacies.

half of respondents having a budget of between £10,000 and £50,000 per year. The numbers of registered disabled students at respondents' institutions ranged from none to 10,000, with less than half (57) having between 100 and 1,000 registered disabled students.

(a) Audit Cultures and Accounting Imperatives

> Aggregate characteristics of the population had to be available in order to formulate policies to address them. A new discourse – statistics – evolved to meet this need, and this discourse became central to the evolving human sciences. The technology of statistics made it possible for governments to create a reality – the statistical facts of their populations – that they could then control (Hekman, 2002:202).

This section of the chapter examines the medical definition of disability employed by SENDA and places it in the managerial context of auditing and accounting. As evidenced from the survey sample, the respondents are employed by a range of institutions in England and Wales. Across this diverse group of disability officers, there is widespread concern about the ways in which SENDA is constituted, and some respondents feel that the legislation itself is limiting, reductionist and even discriminatory. For example, 61 respondents (52 per cent) noted concern about SENDA's definition of disability. As one disability officer said:

> Although there is a great need for this legislation, it only works when people are fitted into categories and boxes. By applying, it also discriminates.

Another respondent noted:

> SENDA should address social practices rather than specific types of disability [sic].

While a third disability officer said:

> 'Normal day-to-day activity' should include study activities if the person is a student.

Authors and activists in disability studies agree with this view, and have criticised the segregation of disabled people which can result from SENDA (Holloway, 2001; Konur, 2000). The above quotes illustrate two main points of critique: the segregation of disabled people and how this fits into wider cultures of audit. There is a history of categorisation, segregation and exclusion of disabled people in all spheres of life including education (Benjamin, 2002; Race, 2002).[16] One respondent noted the irony of having legislation which is meant to include disabled students, but which could lead to the opposite:

> While SENDA was brought about to create inclusion ... institutions sometimes separate disabled students from the rest, in order to give them special provisions.

If HEIs, in attempting to comply with SENDA, are drawing more attention to students' impairments by separating/segregating them from able-bodied students then they risk increasing their isolation and stigmatisation.[17] Indeed, for a law which is meant to lead to the greater inclusion of disabled students, the definitions it employs are problematic and assume medical models of disability.[18] As one respondent noted about SENDA,

> I think its major shortcomings are (1) its 'medical model' definition of disability.

Another person noted that,

> The philosophy underpinning SENDA should be more clearly based on the social model of disability.

As outlined in Chapter 2, medical models of disability are those which primarily locate disability in individuals, rather than viewing societal norms and practices as disabling (Barnes et al., 1999; Imrie, 2000; Morris, 1991).[19] Another disability officer noted that their institution was trying to move

> beyond the narrow definition of disability as outlined in SENDA.

These concerns about definition are not just academic – legal definitions can materially impact disabled people attempting to access and participate in higher education (Race, 2002). The segregation of disabled people, for example, can be a result of medical definitions and disabling language[20] (Oliver, 1990). Another example was provided by a survey respondent, who felt that SENDA's definition of disability may exclude some people from entering higher education or prevent them from receiving provisions associated with SENDA. This disability officer noted:

---

16   For more on the histories of exclusion of disabled people in education, please see Chapter 5.

17   For a detailed history of "special needs" segregation, please see Chapter 5.

18   The DDA's legal definitions of disability are covered in detail in Chapters 2 and 3.

19   For more on medical versus social models of disability, please see Chapter 2.

20   For more on how language can be tied to material practices and outcomes, please see Chapters 2 and 5.

> The definition of disability is not as encompassing as that of the Further Education Funding Council (FEFCE) or Learning Skills Council. The DDA will actually narrow the definition – students receiving support under FEFCE[21] could be excluded by SENDA.

This view is shared by another respondent, who said:

> There are too many get out clauses e.g. definition of disability may exclude many people who are able to carry out day to day tasks, but not to study at higher level.

As noted in Chapter 2, the idea of measuring people's abilities, based on ideas of normal day-to-day tasks, is a normalising one. The disability officer's comment about 'categories and boxes' shows how SENDA's medical definitions fit into wider managerial cultures of accounting and audit. These cultures broadly include the accounting of all contingencies in HEIs, including costs and benefits, applying quantitative measures to curricula and learning, and taking risk-management approaches to legislation (Barnett, 1990). As one disability officer said:

> The wording used (in SENDA) is often vague and unquantifiable, e.g. normal, substantial, etc. Institutions want quantifiable measures.

As part of this accounting framework, HEIs, like other public sector organisations, have seen the proliferation of audit cultures. Budgets have become itemised, services are being streamlined, and statistics on students, staff, and services are collected and measured in various ways for important reports for use within institutions and for submission to funding councils, Research Excellence Framework (REF) reviews, and the Quality Assurance Agency for Higher Education (QAA).[22] Due to the greater number of ways in which the government can influence them, institutions are becoming more concerned about producing the right reports, complying with regulations, and being able to quantitatively assess and measure results. Educational and non-educational outcomes of students attending HEIs are sometimes perceived as secondary in importance to being able to demonstrate the best results in institutional reporting (Barnett, 2002). As one respondent noted:

> Teaching staff are short of time and feel this (SENDA) is an 'add on'. They are already in some cases spending more time reporting on and documenting their teaching activities than their hours of student contact.

---

21   FEFCE is the now-defunct Further Education Funding Council of England, which has been replaced by the Learning Skills Council.

22   The Quality Assurance Agency (QAA) for Higher Education's code of practice includes comprehensive guidance on ensuring equal access for disabled students.

One example of this is the manner in which REFs are conducted. Funds to support research are allocated in block grants to HEIs but aggregated from several calculations for each department in which there are active research staff.[23] HEIs, and their departments, must find ways of increasing all of these components of the equation in order to maximise their funds. This represents a fundamental shift from the relative autonomy with which HEIs were previously funded, towards a more specific, measurable method of assessing the quality and quantity of research production, and is one indication of how managerialism has altered the HE sector (Barnett, 1990). One respondent noted:

> Our institution is focused on meeting targets and being in compliance with QAA best practices; this is what drives management's goals.

HEIs sometimes scramble to provide the best reports possible, and have services in top shape for REF and QAA inspection exercises, but then become complacent until the next round of reviews. Some institutions appear to be treating SENDA in the same way – producing glossy reports and audits but with little or no follow through. As one respondent noted:

> Very often issues and strategies identified are not fully investigated or implemented and therefore some students feel that just lip service is paid to the reports.

In their current form, regulatory processes like the QAA and REF reviews do not necessarily fundamentally change institutions but can lead to the creation of impressive reports, and institutions appearing to meet these regulatory requirements on paper. As one disability officer, in interview, noted:

> I think there is the QAA code on disability but I still think it's an afterthought for us as institutions. They come back to us to make sure it's included as opposed to it being at the core of assessment concerns ... I just sort of think if you separate everything out all the time ... you can just learn the art of 'yes we're doing this and this because we have a disability office.' ... whereas, to me, these things have to be part and parcel of a whole thing. It's part of my more cynical view ... if you separate everything out you can find ways of responding to these challenges without fundamentally altering your practice.

Like other publicly funded bodies such as local authorities and the NHS, HEIs have learned the art of meeting targets and thus appearing to comply with

---

23   The funds in England are allocated on the basis of this formula: $A=U(Q-1)N$, where A is the sum allocated to an institution for a particular unit, Q is the quality rating, N is the number of research-active academic staff submitted, and U is the basic allocation per point for any particular unit (Dopson and McNay, 1996).

regulations (Newman, 2001). For example, one disability officer noted that in one of the buildings at their institution, the lifts often do not work, so that some people with physical impairments cannot access classrooms. Of even more concern are fire egress plans:

> On paper our health and safety [issues] are sorted. But I shudder to think what would happen if there is a fire and we have students in wheelchairs in the building. There are plans on paper, but I'm not sure how feasible they are. We do not want surprise visits from the Health and Safety Executive.

This is what Schuck (1999) calls a 'limit of the law', i.e. there is legislation in place, the actors responsible are aware of it and are in compliance on paper, but perhaps not in reality. Legal regulations alone cannot change this. Law is dependent on people, and laws like SENDA rely on the goodwill of institutional actors to make their HEIs inclusive (Blomley et al., 2001). In cultures of audit, increasing emphasis is placed on performance indicators, service targets, accountability and reporting of data (Power, 1999). As one survey respondent said:

> The Principal and senior management are responsive (to SENDA) but some lecturers need more guidance in their response as do some department heads. If they are not accountable for it, it is not on their 'radar', it does not register as an issue.

Much of the language of accounting and audit appears to have been imported into higher education from the private sector (Barnett, 2002). For example, servicing the economy, by providing graduates for the job market, is seen as more valuable than the creation and transmission of new knowledge (ibid.). As such, there has been a fundamental shift in the way we think about higher education. HEIs are increasingly seen as sources of skilled workers for national and global economies (Scott, 2000). In the context of an increasingly fiscally aware higher education sector, how are HEIs viewing SENDA? There has been apprehension, and even hostility towards SENDA because it is seen as obligating universities to a new set of responsibilities, without ensuring the resources to meet these responsibilities (Palfreyman and Warner, 2002). An example of this hostility is noted by one disability officer who said:

> I face attitude and often genuine ignorance of issues relating to some disabled learners.

In Foucauldian (1979) terms, disabled students with new and unknown needs present a challenge for institutions as they fall outside of what has been classified, categorised and controlled. Using 'normalising' technologies of governmentality, institutional actors feel they must try to 'know' and 'contain' these forms of difference, and control them or else risk chaos and anarchy. Palfreyman and

Warner's (2002:423) quote illustrates this very real and palpable fear in the HE sector:

> The particular piece of legislation is not only relatively new, complex and wide-ranging (hence, potentially expensive) in its applicability to HEIs, but it also involves matters of some moral force (if not also ones of great sensitivity in terms of "political correctness") and is evolving further. The HEI manager needs to tread carefully.

If this is the way HEI managers are being advised to respond to the DDA, what are the chances of disabled people achieving equality within institutions? Moreover, the view taken by Palfreyman and Warner (2002), who are widely read in the sector, elicits scepticism, and perhaps even fear, in HEI managers. The main apparent concerns are with cost, burden of responsibility, and institutional image. Even the reference to moral issues is framed in terms of 'political correctness', which is hardly the pinnacle of inclusion and equality (Young, 1990). Survey respondents echoed this sceptical point of view about SENDA. When asked if institutions are happy with SENDA, some of the responses included:

> No – cost, cost, cost.

Another person noted the

> lack of awareness regarding specific tangible responsibilities, lack of ring-fenced disability funding, fear of litigation.

One respondent said their institution's less than positive response to SENDA is because of

> fear of the unknown

while another respondent said of their HEI,

> they see it as another set of responsibilities with no benefit to the HEI.

These quotes illustrate the unease created in institutions when students and their potential actions (including legal ones), cannot be fully known, understood or accounted for. This echoes Foucault's (1979) notions that power is diffuse and everywhere – it is not just in the hands of senior institutional managers. It is also clear in many institutions that inclusive attitudes have not penetrated or filtered down through these multiple and complex layers. What SENDA and managerialist conceptions of audit and accounting fail to consider are the importance of local practices, knowledge, and institutional variation.[24] I found much variation in institutional attitudes towards, and provision for, disabled students, in my surveys. For example, budgets for disability services range from £1000s to £1,000,000s (see Figure 4.1).

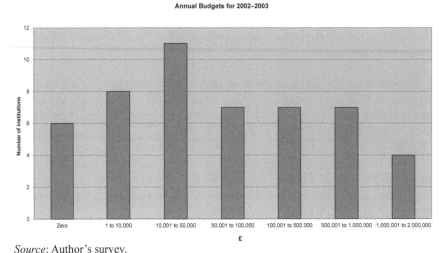

Annual Budgets for 2002–2003

Source: Author's survey.

Figure 4.1   What is your office's annual budget (for academic year 2002–03)?

This indicates huge variation depending on local resources and priorities. While it is accepted that making institutions accessible will require significant resources, the majority of HEIs did not increase the budgets of their disability offices in response to SENDA. Again, this is in the context of an increasingly managerialist HE sector in which institutions are expected to do more with less. As Table 4.1 indicates, the budgets of disability offices have not grown in anticipation of SENDA.

In a competitive higher education sector, HEIs have changed their institutional practices, both in terms of teaching and learning, as well as administrative functions. For example, lecturers are more valued if they can attract research grants to institutions. University departments increasingly have to justify budgetary expenditures, and, since the introduction of tuition fees in 1998, top-up fees in 2006, and the trebling of fees in 2011, they are responsible for the generation and management of their income from students. Students have also changed, and because they are paying for their education, some argue that they are increasingly being cast as 'consumers' (Barnett, 2002). As Palfreyman and Warner (2002:4) note, "higher education in the UK is a big business in every sense of the word, and it is growing." They note that in England and Wales, funding councils directly fund 153 HEIs and a large number of FE institutions for their HE work. The sector

---

24   For more on the importance of local practices and institution variation, see Chapters 5 and 6.

Table 4.1 **Has your budget grown for this coming academic year (2002–03) in anticipation of the DDA and Special Educational Needs and Disability Act (SENDA)?**

Response	Number (%)
Yes	39 (34)
No	52 (45)
Not yet known	5 (3.8)
Not applicable/blank	19 (15.8)
Unlikely	1 (0.8)
I have never been refused funding	1 (0.8)
Other[*]	1 (0.8)
TOTAL	118 (100)

[*] One respondent answered 'other' because her line manager kept the budget secret from her.

Source: Author's survey.

employs 250,000 staff, and teaches more than 1.5 million students in the UK annually, and the sector's annual turnover is said to be over £10 billion (ibid.).

If HE is a big business, why have not more institutional budgets for disability offices increased? Some HEIs are employing a risk management strategy i.e. allocating the minimal amount of investment required to avoid disability litigation, as the likelihood of students taking legal action is considered small. There has been very little case law to date, with only three DRC-supported cases between 2002, when SENDA came into force, and the establishment of the new Equality and Human Rights Commission (EHRC) in 2007, which replaced the DRC.[25] Juxtaposing the responses from Table 4.2 with what disability officers consider areas of challenge, it seems that while HEIs are pressured to become more accessible in a number of areas, they are not increasing their resources in these areas. These challenges – what disability officers consider the main problems – are highlighted in Table 4.2.

As this table indicates, many students face severe and major problems in accessing institutions and their learning services. But a majority of respondents said their HEIs had not increased their budgets in response to SENDA. How then can inclusion be achieved? This shortfall also shows how antidiscrimination

---

25   Indeed, one solicitor who takes SENDA cases said that she generally advises clients not to pursue legal cases as they are time-consuming, costly and not guaranteed to provide positive outcomes for students. For example, in the case of DRC/03/7585, the HEI had to pay £2,500 for injury to the student's feelings. This figure represents a very small amount of most institution's budgets and in some cases may be cheaper than making anticipatory adjustments. This is just one example of taking a 'risk management' approach. See footnote 30 for more detail on case law.

Table 4.2   What are the main problems confronting disabled students' access to buildings and services at your institutions?

Problems faced by disabled students	Severe problem Number (%)	Major problem Number (%)	Minor problem Number (%)	No problem Number (%)
Access to the university campus itself	8 (5.7)	11 (9)	52 (44)	38 (32)
Transport to the campus	12 (10)	24 (20)	40 (33.8)	28 (23.7)
Access into buildings	13 (11)	19 (16)	60 (50)	14 (11.8)
Stairs and lifts	16 (13.5)	25 (21)	51 (43)	14 (11.8)
Signage	9 (7)	32 (27)	45 (38)	19 (16)
Hearing interpreters	31 (26)	18 (15)	24 (20)	31 (26)
Braille copies of notes	9 (7)	15 (12.7)	31 (26)	40 (33.8)
Providing materials for dyslexic students	3 (2.5)	5 (4)	47 (39.8)	48 (40)
Accessible toilets	4 (3)	9 (7.6)	54 (45.7)	43 (36)
Prejudicial attitudes	2 (1.5)	17 (14)	63 (53)	18 (15)

Source: Author's survey.

law can only go so far; without adequate resources, no legislation will eliminate the problems faced by disabled students. It also indicates, in keeping with new managerialism, that resources and budgets are key sources of control and power within organisations. More important than legal and policy stances are the ways in which organisations allocate funding.[26] Financial control appears to be as if not more critical to creating inclusion than policies mandating HEIs to do so.

Another problem, beyond having access to sufficient budgetary resources, is the difficulty of obtaining non-financial resources such as technical information, hiring, training and retaining appropriate members of staff, and gaining expertise in issues relating to disability. For example, one person highlighted challenges in

staff awareness, national shortage of BSL interpreters, and recruitment of personal assistants.

---

26   For more on this, please see Chapter 5.

In all of the surveys and interviews, resources seem to be the key factor in determining an institution's access strategy. Non-financial resources such as British Sign Language (BSL) interpreters cannot simply be bought, but require new practices, skills and knowledge bases. As Barnett (2002) notes, the free market, as applied to the HE sector, does not always 'take care' of everything.

This section has indicated how SENDA's medicalised definitions enmesh with increasing tendencies to audit all aspects of institutional life in HEIs. The next section examines the new managerialism's tendency to centralise power and how this emanates in relation to SENDA.

(b) The Managerialisation of Local Actors

> The DDA 1995 is an example of the kind of increasingly complex and ever-encroaching legislation which tends to creep up on the unsuspecting manager of an HEI (Palfreyman and Warner, 2002:423).

The above quote indicates the kind of advice given to managers in the higher education sector, as Palfreyman and Warner (2002) are lawyers who specialise in educational policy issues. The fact that SENDA can be seen as 'ever-encroaching', and the idea of a piece of legislation 'creeping up' on managers, indicates the uncertainty, fear and even negativity with which HEIs are responding to it. While some of these responses may stem from ableist attitudes, and a reluctance to act on issues of inclusion, they can also be seen in the context of the increasing role of the state in managing higher education and the managerialisation and centralisation taking place within HEIs.

Building on the last section's analysis of the use of accounting and auditing frameworks in HE, this section examines these centralising processes as evidence of managerialism at work in HEIs responding to SENDA. These managerial processes, employing the medicalised definitions of disability discussed in the last section, indicate particular ways of thinking about disabled people and the use of prescriptive measures to address their needs. For example, one feature of the new managerialism is its concern with and anxiety around issues of legislative compliance. As one disability officer noted:

> Disability officers are happy that these issues now have to be taken up at institutional level but other members see it as another burden and are anxious about legislation.

Managerialisation in England has seen businesses and institutions generally being cautious and anxious about regulatory and administrative dictates from state agencies, for example, Health and Safety laws, the DDA Part 3, etc. (Newman, 2001). In HEIs, there is a direct linking of government funding, institutional prestige and perceptions of quality in relation to the RAE, QAA and HEFCE (Barnett, 2002). HEIs have primarily responded to these diktats by creating more rigid reporting structures and allowing for less institutional flexibility and autonomy. Indeed, a feature of the new managerialism has been the tendency to challenge traditional practices of institutional life. For example, a survey respondent noted:

> I think welfare officers are (happy) but SENDA has challenged traditional methods/practices for more established staff who belong to the 'old school.'

This terminology indicates that some institutions still view services provided to disabled students in terms of 'welfare', a term associated with ableist, paternalistic notions of the state (Stone, 1985). The above quote also reveals challenges to 'old school' and old managerial notions. As Deem (1998:47) notes, old styles of governing institutional life were not even considered management, and to use the term management "would have been regarded as heretical." Universities were thought to be autonomous, self-governing collectives of collegial scholars working together, teaching students and learning from each other (Barnett, 1990; Deem, 1998; Kogan and Hanney, 2000). Those individuals responsible for governing institutions were seen as academic leaders rather than managers.

While these challenges are positive – in as far as 'old' management was an ableist, upper class, male domain – new managerialism seeks to replace these old ideas not with practices which are inclusive, but with policies and procedures which fulfil legislative requirements and maximise notions of compliance and covering all legal bases (Gooding, 2000; hooks, 2003). As outlined in the previous section, cultures of audit and accounting have been absorbed into HE. These cultures have material impacts on the day-to-day work, policies and procedures carried out by disability officers. This can influence services for disabled students, for example, with the regimented procedures used to 'diagnose' students with dyslexia and other learning impairments.[27] As Deem (1998:48) notes, the 'explicit' and 'overt' management of staff in HEIs is also becoming more common. Most disability officers who responded to the survey – whilst committed to inclusion – are demoralised, over-worked or over-managed. For example, one disability officer said about SENDA,

> A lot of time it's seen as added work with no added resources being given by management.

Another survey respondent noted:

> I believe that many institutions view SENDA as just another government initiative that results in more work.

Many disability officers echoed this view, i.e. that they personally respond positively to SENDA, but their institutions see it as an example of the government

---

27   For more on issues of 'diagnosis' of impairment please see Chapter 6.

'encroaching' upon them, in Palfreyman and Warner's (2002) lexicon. Foucault's (1983) notion of pastoral power illustrates two issues in relation to this idea (in Dreyfus and Rabinow, 1983). The first is that the state, through the functions of higher education, is spreading the reach of its power into non-state institutions in subtle and diffuse ways (ibid.). The second is that disability officers, as actors in these matrixes of pastoral power, are indeed becoming locuses of that power. They are seen as institutions' 'eyes' and 'ears' on disability issues (disability officer in interview), are expected to manage responses to SENDA, and are accountable for meeting the needs of disabled students, as evidenced in the previous section (ibid.).

Given the varying views of disability officers and their HEIs, overall institutional responses to SENDA are ambivalent, and in many cases, negative.[28] For example, in Table 4.3, only 23 per cent of respondents say that universities are happy with SENDA. This is significant, because it influences the actions that are taken in response to SENDA.

Table 4.3   In your experience, are universities generally happy with the DDA/SENDA?

Yes Number (%)	No Number (%)	Don't Know Number (%)	Blank Number (%)	Total Number (%)
31 (23)	15 (13)	67 (57)	5 (4)	118 (100)

Source: Author's survey.

In addition to the 23 per cent of respondents who say that universities are happy with SENDA, 13 per cent say they are not happy with the law, and 57 per cent say they do not know whether universities are happy or not. While some institutions have embraced SENDA, there are others who see it as burdensome, yet another element of the government's *Widening Participation* agenda for higher education. The reality is that there has not been a direct linking of SENDA and the Office for Fair Access (Offa), the government's *White Papers* on higher education, or the *Widening Participation* agenda. Regardless of the reality, the perception is that SENDA, along with these other initiatives, is a further example of the government's desire to control the higher education sector (Barnett, 2002; Palfreyman and Warner, 2002).

The government's various mechanisms of intervening into the higher education sector can be seen as examples of power's "capillary forms of existence", spreading from the state into institutions and individuals, and inducing them to regulate themselves (Foucault, 1980:39). In addition to the centralising tendencies of the government, there is a similar centralising process occurring in many HEIs.

---

28   There exists, of course, huge variation in responses, depending on institutional type. For more on the importance of institutional specificity, see Chapter 5.

For example, one disability officer said that senior managers had taken more control of their activities. About SENDA, the disability officer noted:

> It may curtail some of our activities because of fear of litigation.

While fear was mentioned in section (a), I now use this concept to demonstrate how this fear impacts on actions of institutional actors, particularly senior managers. Many disability officers noted how their institutions were undergoing similar processes of tightening, restricting and regulating of their activities in response to SENDA. This echoes Foucault's governmentality, a mentality or ethos by which senior managers employ elaborate methods and technologies to categorise, classify and control more aspects of institutional life (Foucault, in Dreyfus and Rabinow, 1983). This leads to more centralised control of resources within institutions and can mean that disability officers often do not have the power to commit the effort or resources necessary to respond appropriately to SENDA. As Kogan (1999:263–264) says:

> Institutions respond to external changes and their responses become structurated in terms of organisational power structures. Many of the more important changes have been described as bureaucratisation ... the move from individual and academic power within the often mythic collegium to the system or institution, and a resulting new structuration of decision-making.

Disability officers who would like to see their institutions become more inclusive often do not have the authority to make such changes. Some institutions are responding to SENDA by attempting to centralise power and maintain tighter control over policies and procedures.[29] This also mirrors Foucault's (1979) notion of institutions which attempt to spread their disciplining forces right down to all levels of activity. The most significant decision-making members of HEIs are senior managers, and as the survey shows, their responses have not necessarily been positive. More senior managers were 'not responsive' to SENDA than any other group. As Table 4.4 indicates, different categories of staff responded differently to SENDA's provisions, which means there is no way of guaranteeing uniform actions in response to SENDA within institutions. For example, when asked if they faced more challenges in getting particular groups of colleagues at their institution to respond to the DDA, 40 per cent replied "yes," with most respondents citing either senior managers or academic staff as being the least responsive. One person said that their senior manager

> Is not taking the issue seriously or allocating funding, and some lecturers are not accepting responsibility to make curricula accessible.

---

29   Some of this has to do with public relations concerns and HEIs avoiding negative press. For more detailed discussion of this issue, please see Chapter 5.

Table 4.4   **How responsive are colleagues at your university to the provisions of the DDA/SENDA?**[30]

Categories of colleagues	Towards all aspects Number (%)	Towards most aspects Number (%)	Towards some aspects Number (%)	Not responsive Number (%)	Hostile Number (%)
Department Chairs/ Heads	11 (9)	60 (51)	29 (25)	2 (1.6)	1 (0.8)
Lecturers, Readers and Professors	6 (5)	54 (46)	41 (35)	2 (1.6)	
Admin. Staff	11 (9)	64 (54)	28 (24)	1 (0.8)	
Senior Managers (Vice-Chancellors/ Principals)	16 (14)	55 (47)	24 (20)	5 (4)	1 (0.8)

Source: Author's survey.

Another person noted that,

> I am still trying to convince the senior managers that the DDA is a serious legal instrument and we must take it on.

This indicates that even in cases in which disability officers are responsive to the DDA, they do not always succeed in eliciting wider institutional responses to it, especially from those in positions of authority. The task of motivating their institutions to meaningfully engage with this legislation is not a straightforward one, even for the disability officer who is committed to doing so. Keeping Foucault (1979) in mind, all institutional actors have power, but that power exists in relation to the power of other actors. If those actors have ableist attitudes, or are driven by managerialist values, these will mitigate disability officers' abilities to exert their power in favour of creating inclusion for disabled students.

In their new managerialist modes, HEIs have become more hierarchical and top-heavy, and this can influence the inclusion or exclusion of disabled students, for example, from institutional governance and decision-making processes. Indeed, some institutions have restricted policy-making to senior management teams alone, and have foreclosed disability officers' access to information, in attempts to consolidate power in their decision-making processes. As one respondent said:

---

30   Percentages have been rounded up to the nearest decimal.

> Disabled students were represented on my committee but since change of line management in 2001, student representatives stopped being invited. Since July/ August 2002 I am not invited to meetings either.

Another respondent noted the opposite – a completely decentralised mode of policy-making at their institution, and said:

> University policy is not driven from the centre. We will have to take each aspect and work with people to start developing their own strategies to deal with their own particular area of expertise. If we tried to dictate we would get no co-operation whatsoever.[31]

Simultaneously to the centralising processes occurring in some HEIs, the responsibility for all disability issues often rests primarily with disability officers. This means disability officers may have responsibility for disability issues, whilst having no authority or power to make significant changes.[32] The people 'on the ground', i.e. disability officers, are often required to ensure that the institution is meeting its SENDA requirements, without the necessary resources, knowledge or staff to do so. For example, 27 disability officers surveyed (23 per cent) had no training on disability issues. As Figure 4.2 demonstrates, even those that had training received it using a diverse set of organisations and documents. Only 11 respondents (9 per cent) had DDA-specific training. This indicates that there cannot be uniformity in all HEIs approaches to SENDA.

One problematic aspect that Figure 4.2 reveals is the strong reliance on the national disability charities for training and expertise. Many disability studies writers take issue with these charities for their patronising campaigns and paternalistic modes of operation (Morris, 1991; Oliver and Barnes, 1998). Specifically, they criticise disability charities for being run primarily by able-bodied people, thus appropriating disabled people's voices and experiences into catchy sound-bites or pity-seeking advertisements in order to raise funds, and perpetuating paternalistic attitudes towards disabled people (Clare, 1999; Morris, 1991).

Disability officers also face a number of concerns, including raising awareness among members of staff who deal with disabled students on a frequent basis. As one respondent said:

> There is not enough specific training and funding attached to enable establishments to respond. We struggle with support for blind students because of difficulty finding suitable resources.

---

31   Such institutions are now a minority, and mainly fall within the category of Russell Group universities, many of which comprise smaller colleges within the larger institution.

32   For more literature on disability officers as 'lone champions' of equal opportunities' issues in their institutions, see Chapter 5.

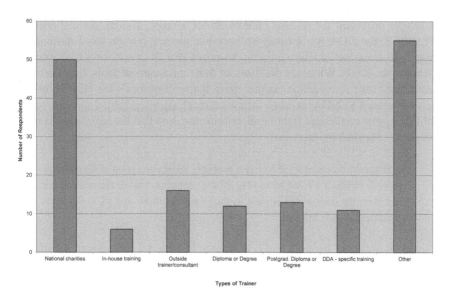

Source: Author's survey.

Figure 4.2   If you did receive training, can you say what kind of training and who provided it?

Another disability officer noted:

> Regarding provision for carers ... finding qualified people is a problem.

Even in cases where funding is available, finding qualified people appears to be a challenge. One person felt that

> It's always a worry with high turnover of part-time staff that all staff are familiar with the implications of the Act.

Another person noted the difficulty in receiving proper training and said that disability

> is not seen as an issue at management level.

One disability officer, on a more positive note, said:

> (SENDA) can force senior management response where there has in the past been an opt-out opportunity.

While this indicates a positive response to the legislation, SENDA has been criticised for its lack of enforcement procedures. Unlike the QAA or RAE exercises, it has no provision for reviews, follow-up, or a rating system. As discussed in Chapter 2, the DDA is not easily enforceable, as they rely on the discriminated parties to bring actions forward, and to prove that discrimination has occurred (Roulstone, 2003). Whatever the flaws of these enforcement tools, they do make institutions take some action, as the above quote illustrates. One HEFCE officer felt that even if SENDA had such follow-up mechanisms, disabled students would still face great challenges fighting discrimination, and that the legislation is not a panacea.[33] He said, in interview:

> I think the weakness of it [SENDA] is probably about perception at the moment. Everyone thinks it's going to resolve all the problems. I think the weakness is that it's on the onus of the student to take the institutions and all the inherent difficulties that go with that.

Finally, as noted in Chapter 2, and as illustrated in Table 4.4, attitudes are significant to SENDA's impacts on disabled students' lives. Attitudes about disability range from a "frosty response" to SENDA among colleagues to "the new Vice-Chancellor is very positive." One respondent said, of SENDA, "they don't understand it" and have "poor attitudes towards disabled students." As one person noted:

> Some academics are worried about workloads and others are concerned about cost. Universities are at different stages, some built up services in the 90s with HEFCE project funding, others did not. Levels of awareness also fluctuate from place to place.

The survey responses indicate a huge variation of views, attitudes and responses to the DDA and SENDA, both in terms of disability officers' own views of the legislation, and from their representations of their institutional engagement with SENDA. The openness and receptivity to SENDA appears to vary at different institutions, based on survey responses, from hostile, to very inclusive, to indifferent. One respondent noted that

> academic staff have demonstrated very mixed reaction to SENDA.

---

33   There have been few SENDA cases to date. Some argue that it is very difficult for students to prove discrimination (Mann, 2002). For example, the DRC has supported only two cases of students proving less favourable treatment in relation to HE (DRC/03/8637 and DRC/03/8629, both contesting their lack of admission into nursing programs; the latter is still pending an outcome), and one case of a student proving that he was refused reasonable adjustments (DRC/03/7585, a student with 'mental health issues' who requested separate accommodation in a quiet area). For more detail, see www.drc-gb.org

In this section I have illustrated the diverse ways in which processes of managerialisation and centralisation influence disability officers' abilities to implement SENDA. We have seen the challenges – material and attitudinal – faced by disability officers trying to change institutions which are struggling to preserve traditions whilst introducing increasing managerialist modes of operation. As evidenced by some of the quotes, SENDA can have impacts in terms of making institutions fear lawsuits and negative publicity, but because of the large numbers and variety of staff and lecturers in HEIs, the law's grasp cannot reach all institutional actors, and so it is up to individuals to change their attitudes and actions. The next and final section concludes the chapter.

(4.5) Conclusions

In relation to Foucault's (1979) theories that power is everywhere, disability officers, in some instances, feel powerless and over-managed. Overall, regardless of managerialism, or perhaps because of it, HEIs are undergoing authoritarian 'clamping down' on the actions of individual actors such as disability officers. As Radcliffe (1999:237) notes,

> The practices of power – the quotidian operations to place objects and subjectivities in relation to each other – operate unevenly over space. Such uneven processes and results arise from the engagement of social relations with the spatial and located nature of power relations.

Despite SENDA's power to make disability discrimination illegal, this chapter has illustrated that changing day-to-day practices in HEIs requires much more than legal remedies. Disability officers need budgets, training, adequately staffed offices, and the authority with which to make real changes in their institutions. These are the material resources which SENDA does not provide. Disability officers are struggling to keep up with increasing managerialist demands for efficiency, value-for-money, and more centralised and bureaucratised control over their actions and functions.

Apart from lacking material support structures to accompany its legal requirements, SENDA also demonstrates that law alone cannot change attitudes. Many disability officers reported negative attitudes towards disabled students, from lecturers, senior managers and other staff, despite the existence of the law. With its weak enforcement tools, and the perception that it is difficult to locate discriminatory attitudes and behaviours, SENDA relies too heavily on the goodwill and actions of individuals. My postal survey indicates that responses to SENDA depend on many different factors – institutional type, institutional history and ethos of inclusion, and the powers and personalities of local actors. Ideas about institutions, and the variations between and among them, will be further explored in Chapter 5.

Chapter 5
Accessing the Ivory Tower:
The Importance of Institutional Geographies

(5.0) Introduction

The first universities were founded as temples of higher learning but more recently, particularly since the introduction of professional qualifications in universities, there has been an impassioned debate about the purpose of university education (Bourdieu, in Savage et al., 1992; Makdisi, 1984). As Twining notes: "from outside the ivory tower there has nearly always been pressure ... to make university education more obviously useful and vocational" (in Barrett, 1998:145). Taking a firm stance against the vocational nature of some university degree programmes, he argues that "a university is not a trade school for the production of plumbers" (ibid.).

Twining's pronouncement illustrates the historical and ongoing tensions around access to higher education, and the desire to produce a societal elite versus liberal egalitarian values (Bourdieu, in Savage, et al., 1992). This chapter builds on the analysis of the Higher Education (HE) sector in Chapter 4. I consider how institutions are being affected by, and responding to, SENDA.[1] While Chapter 4 employed a sector-wide analysis, this one examines HE at the institutional scale. In geographical research, analysing the subject matter at different scales is important for both empirical and epistemological reasons (Blomley et al., 2001). As noted in Chapter 1, this book analyses the impact of disability discrimination legislation across various geographical scales – from the level of the nation-state to the daily, lived experiences of disabled students. In between are various layers, including service providers in the UK, a sample of surveyed HEIs[2] in England and Wales, and, in this chapter, two specific institutions selected to illustrate how understandings of the DDA are unfolding in the UK higher educational context.

This chapter provides insights into how HEIs are responding to SENDA, and how these different ways of responding are mediated by axiomatic differences and institution-specific factors. This relates to a central geographical argument that difference and specificity are important. For example, Massey (1997:66) points out the "highly complex social differentiation" within places and argues that

---

1   The DDA consists of Parts I through V. The Special Educational Needs and Disability Act (SENDA) is Part IV of the DDA and applies to higher education institutions. For the purposes of this chapter Part IV will be identified as SENDA.

2   HEIs include universities, higher education colleges and further education colleges for the purposes of SENDA, and when I refer to HEIs I include all of the above.

"places do not have single, unique 'identities'; they are full of internal conflicts. It is therefore important to interrogate place for the purpose of assessing how law interacts with the institutional spaces impacted by the DDA in varied and multitudinal ways.

By focusing on *the particular*, this chapter considers how law does or does not take geographical specificity into account, even as it aims to create reasonable adjustments for disabled people, who by legal definition, are different.[3] As noted in Chapter 1, a number of critical legal scholars have challenged the view that law is rational, abstract, and objective (Blomley, 1994; Clark, 2001; Pue, 1990). Blomley (1994:11) says, "critics argue that, far from constituting an autonomous sphere, the legal project is necessarily a social and political project". In examining how SENDA is being interpreted and implemented at an institutional level, it is important to contest the abstract neutrality of law because critical legal scholars posit that law is relational, that is, it acquires meaning through social action (ibid.).

This chapter examines how the law, exemplified by SENDA, acquires meaning through the social actions, rules and regulations in the day-to-day operations of two institutions. The chapter is divided into five sections. Section (5.1) introduces the chapter's broad themes which centre on unearthing liberal values in higher education. Second, (5.2) provides a theoretical framework for examining institutions. Third, (5.3) compares and contrasts two institutions and argue that an understanding of the implications of SENDA ought to be related, in part, to the social, cultural and geographical characteristics of the respective institutions. Section (5.4) explores and evaluates the contrasting ways in which institutional specificity influences the understanding of disability and the needs of disabled students. As evidence indicates, the legal complexity of SENDA has been compounded by the institutional complexity of HEIs and the challenges of implementing SENDA in very different institutional contexts (Konur, 2001). Section (5.5) describes and discusses the contrasting responses of the two case institutions to the needs of disabled students in relation to their respective interpretations of SENDA. Section (5.6) concludes by suggesting that a more thorough understanding of how HEIs operate needs to be included in legal discourses in order for SENDA to achieve its goals.

(5.1) Unravelling Liberal Values in Higher Education

As a central theme in this chapter, I examine the concept of values in order to unravel, demystify and expose institutional attitudes and practices towards disability and disabled students. As Williams (1997a:15) notes, examining the meaning of HE requires considering "not simply the language through which it is

---

3   DDA definition of disabled person. The Act defines a disabled person as someone with "a physical or mental impairment which has a substantial and long-term adverse effect on his ability to carry out normal day-to-day activities." (HMSO, 1995).

presented but also the institutional practices and the social and political positions of those who 'speak' about higher education."

Moreover, the values of a HEI influence not only the content of educational curriculum but also the broader discourses and everyday practices which form institutional life (Williams, 1997a). Examining institutional values is critical to understanding HEIs as these values inform the overt and visible forms of institutional life – what can be seen, heard, experienced and felt (Barnett, 1990). Additionally, values also interact with more subtle but powerful realities in institutions: they can influence key players, policies and practices; institutions operate in both formal and informal ways, with the latter being more difficult to trace (Ball, 1990).

Finally, it is important to unravel values to demonstrate that they are not necessarily a panacea for changing institutional discourses. As Foucault (1979) notes, liberal values can be used in institutions to maintain and reinforce dominant discourses and hierarchical power relations rather than to subvert them. Therefore, accepting the institutions' commitments to liberal egalitarian values prima facie carries with it the risk of allowing inequality to continue. For example, in her book *Academic Women*, Brooks (1997:1) notes that there is a

> clear contradiction between the model of the academic community characterised
> by equality and academic fairness, which academic institutions purport to have,
> and the reality of academic life within these institutions.

In her research, she interviews women in academia about their experiences with a view to unearthing the contradictions between egalitarian values of HEIs versus the lived reality in academia. Brooks (1997:1) finds that this lived reality consists in part of "male-dominated hierarchies which lead to endemic sexism and racism in defence of male privilege." Several writers note that egalitarian values and policies themselves do not automatically lead to equality for all, and an end to sexism, racism, homophobia and ableism in HEIs (Barnett, 2000; Benjamin, 2002; Brooks, 1997; hooks, 2003; Razack, 1998).

What are the challenges faced by institutions in transmitting their egalitarian goals into lived realities? Are these egalitarian values genuine institutional commitments or just attempts at legal compliance and maintaining acceptable public relations? What are the obstacles to creating equality in HEIs? In the context of disability discrimination, this chapter examines (5.4) the attitudes, values and practices around disability in two case study HEIs, and (5.5) how these values are actualised and put into place in response to SENDA, as well as the institutional resistance to these responses. In the next section I provide a theoretical framework before moving on to the institutional findings.

(5.2) Institutions and Institutionalisation: Contingency and Complexity in HEIs

> Without an understanding of how responses to subordinate groups are socially
> organised to sustain existing power arrangements, we cannot hope either to
> communicate across social hierarchies or work to eliminate them (Razack, 1998:8).

I argue that policies, procedures and daily actions in HEIs have specific social
and historical contexts which need to be examined (Barnett, 1990; hooks,
1994; Razack, 1998). Taking HEIs' policy statements at face value – without
examining the deep-rooted values of elitism, exclusion and hierarchy – would
be superficial and could allow the deeply discriminatory histories, and present
day realities of HEIs, to go unchecked (hooks, 1994, 2003). Building on the last
section's discussion of liberal values in HE, in this section I outline the broad
theoretical framework used in the chapter. As mentioned in (5.1), this chapter
critically examines how disability is viewed, and what is being done in response
to SENDA, in two particular HEIs. This examination takes place in the context
of liberal egalitarian ideas of education, and the rights discourses within which
SENDA is situated.[4] My central argument is that institutions matter: *what* types of
institutions, *who* influences them, *how* they are run, and *where* they are located.
These specific factors can and do influence institutional responses to SENDA, and
outcomes for disabled students. To support this argument I draw on a diverse range
of literatures which address the importance of institutional actors, dynamics and
practices, including in the fields of organisational culture, legal geographies, post-
structuralism, and critics of managerialism in higher education (Barnett, 1990;
Blake et al.,1998; Blomley, 1994; Clark, 2001; Forrest, 1998; Foucault, 1979;
Nash and Calonico, 1993).

Evaluating SENDA in the context of institutions is important because it is
within institutions that legal implications exist. SENDA specifies an ideal of
'non-discrimination' and the requirement to make 'reasonable adjustments'
but how are these understood and implemented in HEIs? Institutions and the
differences between and within them are germane to how legislation like SENDA
is implemented (Barnett, 2000). Institutions are important because they are the
sites of discourses through which SENDA is mediated. The idea that SENDA will
'filter' through all HEIs and produce equal outcomes for disabled students ignores
the contingency and specificity of local institutional actors, cultures and socio-
political environments (ibid.).

Forrest (1998:1), for example, writes about a 'new institutionalism' which
pays particular attention to the ways in which administrative processes at the
institutional level can help to reveal broader discourses of society and state, and
"explain how policies are formulated, considered and resolved." He aims to move
away from abstract conceptions of the state towards a more nuanced understanding
of how institutional specificity and difference can influence the outcomes of

---

4   For a further discussion of rights discourses of the DDA, see Chapter 2.

policies in practice (ibid.). Likewise, Foucault (1993) sees the state as not a unified, coherent structure but rather as an "ensemble of institutions, procedures, tactics, calculations, knowledges, and technologies" (in Gane and Johnson, 1993:7). Both Forrest's (1998) and Foucault's (1993) analyses emphasise the importance of the local, the specific and of how the shape and form of institutions and their day to day practices mediate outcomes for the polity.

The intricate power dynamics and complexities of institutions come under the examination of Foucault in his theories of governmentality (1979) and power-knowledge (1980). Discourse is the central concept in Foucault's analytic framework, and he uses this to argue that knowledge can determine meaning and its representations, and that the formulation of a discourse is an exercise of power in that it privileges certain viewpoints over others. For example, Christopher Falzon (1998), in his book *Foucault and Social Dialogue*, draws on Foucault's *Discipline and Punish* (1979), and notes that discourses can be used to regulate and control institutional actions and actors, and that normative categories and standards are often used by institutions as socially acceptable and benign ways of controlling behaviours and actions. Falzon (1998:50) notes that a "key tactic for imposing and maintaining such control is to present these normative categories as being universal, necessary and obligatory." Indeed, liberal values are often used as universal ones, as the idea of universality is tied to liberal notions of equality (Razack; 1991; Young, 1990). Foucault (1979:49) notes that discourses are

> practices that systematically form the objects of which they speak. Discourses are not about objects, they do not identify objects, they constitute them and in the practice of doing so conceal their own intervention.

In his work on governmentality, Foucault (1983) argues that institutions – including those in the business of education – play a significant part in the self-governance of liberal subjects (in Dreyfus and Rabinow, 1983). He posits that this occurs through the construction of discourses that define the purposes and practices of institutions and thereby categorise and classify the subjects and outcomes of these institutions. Building on Foucault's work, Marshall (1990) derives five key processes which can be used to help unravel the complex discourses in education: differentiation, normalisation, institutionalisation, compliance and rationalisation. I here define each of these five processes. Differentiation in institutions can be defined as 'dividing' practices conducted by institutional actors in order to distinguish between different groups of students and communities within and outside institutions, creating distinctions such as student/applicant, undergraduate/ postgraduate, male/female and disabled/able-bodied. As Kenway (1990) notes, differentiation is identified by those procedures, which through classification and categorisation, distribute, contain, manipulate and control people ... giving them an identity which is both social and personal (in Williams, 1997a:19).[5]

An example of how differentiation is linked to identity comes from Blake et al. (1998:106), who emphasise the importance of the regional and community role of HEIs. As they say,

> The commitment to promote wider access is in part a new drive for social cohesion ... the divisive nature of progression to higher education, linked as this so clearly is to social class, is increasingly acknowledged. Thus, the regional and community role of the university has an important bearing not only on matters of economic regeneration and growth but also on matters of social identity.

Higher education is inextricably linked to issues of social mobility, class structure and identity politics. In fact, higher education is by definition exclusionary: it is deemed to be 'higher' than the previous level of education, and creates an insider/ outsider distinction with strict rules for membership: either one is admitted, or not. There are, of course, many debates about the role of HE in aiding social mobility, ranging from Bourdieu's discussions of the transmission and perpetuation of class privilege through education systems to the work of more recent theorists who argue that HE should be a tool for transforming social stratification and existing class structures (hooks, 2002).[6]

The second process, normalisation, can be defined as a "naturalisation of a discourse type" (Fairclough, 1989:92), i.e. processes by which certain institutional actors employ particular phrases and labels so as to appear neutral, "and in the best interests of all concerned" (Williams, 1997a:20). Normalisation therefore centres on what objectives are pursued by those with power, and how these objectives come to be seen as "normal, acceptable and legitimate" (ibid.). As an example, notions of 'good students' or 'highest ability' show how selectivity in admissions processes have been normalised, when they are not infallibly neutral measures.

If normalisation is the process of prioritising institutional objectives, then institutionalisation, the third process, represents the ways in which these objectives become translated into practice. The fourth process, compliance, involves the ways in which institutional actors enforce modes of institutionalisation, thus cementing norms. Williams (1997a:20) provides an example of how these processes work together, when she writes,

> Individual institutions engage in a process of ranking and grading qualifications and examination results and this is encouraged and supported by government agencies concerned with admissions, data collection and funding. The data that

---

5   For more on the categorisation of disabled people in education, please see Chapter 6.
6   For more on the HE policies of New Labour, see Chapter 4.

are produced by such processes can be manipulated statistically, and so appear to provide notions of 'objective merit' to rank order students and institutions. This results in a hierarchical differentiation of universities that is documented and institutionalised through the admissions literature and statistical bulletins.[7]

The fifth process, rationalisation, is about the arguments made to sustain institutional norms and discourses (Marshall, 1990). Using the example of access, rationalisation can be demonstrated by how particular forms of merit are seen to be acceptable and legitimate as measures of higher education potential. I draw on these five processes to exemplify institutional specificity throughout the chapter, and in so doing expose the norms and values informing HEIs responses to SENDA.

Because disabled people have historically been systematically disadvantaged, equality may not be achieved by merely allowing more disabled people into higher education (Doyle, 2000; Gooding, 1994; Morris, 1991). A more inclusive and far-reaching goal would be for disabled people to succeed, in large numbers, in a variety of programmes of study, in various institutions, including those which are most prestigious. The so-called New Universities[8] and Colleges of Further and Higher Education have already been addressing and including disability for comparatively longer than those institutions in the Russell Group,[9] for example (Barnett, 2000).[10] It is important to consider not just the numbers of disabled people entering higher education, but also to examine whether practices are changing, and if so, in what ways are they changing? In this section I have emphasised the importance of studying institutional mediation of broader policy frameworks and unearthing and interrogating their value systems. This notion of examining SENDA through the 'lenses' of institutions will be used throughout the chapter. In the next section, I examine the two case study institutions.

(5.3) Shattering Liberal Myths of Higher Education: Institutional Profiles

In her book *Teaching to Transgress: Education as the Practice of Freedom,* hooks provides her personal view of the institutions in which she has studied and/or

---

7   Indeed, there is a raging debate on the relative merits of league tables. See, for example, Hussey and Smith, *The Trouble with Higher Education* (2009).

8   New Universities were formerly known as Polytechnics. The change came following publication of the Dearing Report. For more on this, see Chapter 4.

9   The Russell Group is an association of research-oriented universities in the UK. Founded in 1994 at a meeting held at the Hotel Russell in London, the group is composed of the Vice-Chancellors/Principals of 19 universities including the Universities of London, Bristol, Edinburgh and Birmingham. For a complete list of member institutions, see www. russellgroup.ac.uk.

10   Even at Russell Group institutions, disabled people have always been present, but not necessarily visible.

worked when she notes that "the more I've been in the academy, the more I think about Foucault's *Discipline and Punish: The Birth of the Prison* and the whole idea of how institutions work" (1994:232). Like Foucault (1979), hooks contests liberal notions that education is about the upliftment of individuals, and draws attention to the repressive, coercive aspects of institutional life: "people have this fantasy (as I did when I was young) of colleges being liberatory institutions, when in fact they're so much like every other institution in our culture in terms of *repression* and *containment*" (ibid.).

Hooks' words are salient as these institutional profiles will reveal the deeply liberal values of the two case study HEIs despite the multiple differences between them. The two HEIs are prime examples of the difference that institutional specificity makes. The first, City of Bexley College, is a regionally known FE and HE college; the other, Cromwell, an internationally known university.[11] The latter has among the most competitive entry requirements in the UK, while the former will admit a far wider range of students. This FE/HE college's mission is about access and openness; a student can attend without GCSEs or A-levels,[12] and can enrol in an HE preparation course, an alternative route into HE. Using these two HEIs as case studies illustrates the significance of geographical specificity, and how SENDA does not fully account for the differences and variability in institutional type.

Barnett (1990) has devised a way of conceptualising philosophies of HE, creating a distinction between the forms of education offered at various HEIs. He notes the historical distinction in HE between the goals of developing students' minds, versus the development of the whole person, the latter an attempt to dispel an epistemological split between the mind and the body in education (ibid.).[13] Barnett (1990) also maps the education of the mind versus the whole person onto a conservative-radical matrix of philosophies of higher education. As he says (1990:191):

> In its conservative formulations, there is a strong belief in objective knowledge having been obtained through the research endeavours of the academic community, and it becomes the task of higher education to transmit it. This

---

11   City of Bexley College and Cromwell University are pseudonyms.

12   A GCSE is a General Certificate of Secondary Education. It is traditionally the first qualification obtained by students at school, at the age of 16 and replaced O-levels in 1988 (Tawney, 1998). Students complete up to six GCSE exams, as a prerequisite to obtaining Advanced level qualifications (known as A-levels) which are required for university entry, although this is changing with the introduction of vocational A-levels and HE foundation courses.

13   Mind-body dualism has been used to exclude disabled people for centuries, e.g. the influence of Descartes – the first principle of Cartesian inquiry is to be sceptical. The root of this scepticism is a very rigid dualism between the thinking mind and the physical world outside (see Maslin, 2001).

leads to a relatively passive learning situation, with the student largely on the receiving end of prepackaged knowledge...In its more radical version, though, a sceptical view of knowledge is taken. Accordingly, the student is expected to come to a personal view of the knowledge encountered, and to take up a critical stance. In its most radical form, this view is also accompanied by the idea that the knowledge so won, through personal involvement by the student, can lead to a higher level of self-empowerment.

Using this matrix, Barnett (1990) says that HEIs can broadly be identified as using either conservative or radical interpretations of liberal education but this is not a binary as the continuum in his quotes indicates. The more conservative approaches to HE are "essentially the product of a stratified society" and are based on "a selective system of higher education" (Barnett, 1990:191). He notes that though this conceptualisation of HE contains "elements of Victorian university orthodoxy", its effects are still felt in contemporary HE (ibid.).

Cromwell University, using Barnett's (1990) matrix, approximates the conservative institution. City of Bexley College, conversely, more closely resembles elements of radical interpretations of HE, which aim to provide a student-centred education whose ultimate goal is the "total transformation and emancipation of the individual student" (Barnett, 1990:191). Institutions like Bexley, which embody more radical philosophies, are more likely to couple theoretical and practical elements (for example they offer a course on Hairdressing), promote a 'lived' appreciation of society, and welcome the perspectives of diverse social groups. Barnett (1990), however, says that 'New Universities' in the UK fall short of being truly radical, and that the only successful examples of radical institutions are found in Eastern Europe. Nonetheless, he distinguishes between the conservative sector's claims to "sustain the wider social and cultural functions of HE" and the New Universities and FE/HE sectors' aims to fulfil the 'service' functions of HE in the UK (p.191).

Research at Cromwell revealed two main tensions regarding institutional responses to disability, which are inter-related: (1) a culture of elitism, superiority, and exclusivity, in terms of academic achievement, financial resources, and socio-political influence, and, (2) a desire to be seen as the best and an underlying focus on preserving and enhancing the institution's image.

Notions of exclusivity and image reflect the increasing trend in which institutions are seen as businesses which need to promote and maintain their 'brand value' and pay attention to public relations in order to attract more students and financial resources (Clarke and Newman, 1997; Kogan, 1999). While Cromwell aims for an elite portion of the HE marketplace, Bexley vies for the special needs niche in the HE market, and focuses on selling itself as being friendly and dedicated to providing "a good service" (staff member, in interview). While Cromwell seems to be an archetypal conservative institution, using Barnett's (1990) typology, Bexley's identity is much less established.

Due to government pressure, universities in the Russell Group, including Cromwell, are attempting to be seen as inclusive of students from underprivileged class backgrounds and are required so to do by the Office for Fair Access (Offa).[14] In its admissions procedures, there is a concerted and very visible effort to admit more students from state schools. This is a change from historical admissions practices which favoured students from independent/public schools. In this Widening Participation[15] policy regime, Cromwell has been criticised both by parents of pupils from independent schools[16] and by education critics who see the new admissions policy as an example of tokenism, and say it does nothing to address broader inequalities at the university, such as power structures or financial, cultural or physical barriers to access.

Bexley and Cromwell, like all other HEIs in England and Wales, are in flux and are transitioning into uncertain territory. The HE top-up fees legislation, which came into effect in 2006, added to this uncertainty for all institutions and this uncertainty will be enhanced in light of the coalition governments' scrapping of the teaching grant and new legislation allowing HEIs to charge up to £9000.[17] One certain effect will be further commercialisation of the HE sector as students perceive a university degree as a product for which they paying £27,000.

Because of how power works at Cromwell, the drive to be inclusive of disabled people will have to come from the top-down; it is a very hierarchical place that respects tradition and formal power arrangements. All the persuasive individuals, of which there are few according to interviewees, cannot change Cromwell's over-arching unease with issues relating to disability. Yet conversely, its system of governance, with colleges functioning as semi-autonomous institutions within the larger university, makes it difficult for senior managers to dictate policies to all members of the institution.

This renders it very different from Bexley – a decentralised, community-orientated, public service- driven college of HE and FE. At Bexley, there are many members of staff who are engaged in making their institution more inclusive in the broadest sense, and these efforts are visible. The language of openness, access, and customer service to students is noticeable in marketing materials such as brochures and websites. Students see Bexley as a path to a vocational career, a

---

14   From September 2012, universties can charge up to £9000 but must provide bursaries for students whose family income is below acceptable thresholds.

15   Widening access and improving participation in higher education are a crucial part of HEFCE's mission and form one of the government's strategic aims. See www.hefce.ac.uk

16   For example, at Bristol University, claims were brought to the CRE that a pupil from Westminster Independent School had been discriminated against because of the school she attended. See Macleod, 2003.

17   In March 2004, the Department for Education and Skills announced the introduction of top-up fees of up to £3,000 per year, starting in September 2006, and re-payable by graduates upon completion of their degrees, based on their income and ability to pay. For more on this see Taylor, 2005.

chance to get into HE through via foundation courses, or as an institution that will meet their particular needs, i.e. students with many different types of impairments attend the college.[18]

Historically, Bexley started as a small college which served the local community and offered adult education, extension courses, and evening classes. Due to increasing numbers of students,[19] Bexley took over and bought out other smaller local colleges which were located in away from the city centre or were no longer financially viable. The impact of Bexley taking over different locations on its institutional identity – and its sense of place – should not be discounted. As Cresswell notes, universities have strong identities when he writes that "universities clearly have a number of more or less established meanings as centers of learning, culture, objectivity, humanistic endeavor and reflection ... going back to the Middle Ages" (2004:36).

In the case of Cromwell, with its grand architecture and insider/outsider structure – access into buildings is discreet, with few signs and hidden entrances into colleges, for example – one can see the validity of Cresswell's argument that "a way of establishing the authority of 'professors' was devised and built into the structure of lecture halls" with their raised podiums for lecturers and clear spatial demarcation between instructor and students (2004:36). However, Cresswell warns against seeing these spaces as unchangeable: "it would be wrong to think of the university as a finished place. The traditional arrangement of furniture in the lecture theatre for instance is frequently ignored" (ibid.). Pushing the idea of what a university can be beyond the need for a physical location, Cresswell says that "even more revolutionary are opportunities by the Internet and 'distance learning' that make formal 'placed' education increasingly redundant. The university, as a place then, is not complete. As argued by Cresswell, and by Massey (1997) all places, in general are never complete; they are evolving.

Bexley, with its amorphous campus structure and its regional growth, is an example of Cresswell's point about universities as places in process, and even Cromwell, despite its sense of place having been defined for much longer than that of Bexley, is a place in process, in part due to SENDA. Today Bexley has five campuses. While each does have remnants of its previous autonomous college culture, none has the complete independence of Cromwell's colleges, which have more traditional religious overtones[20] and which have historically functioned as

---

18   For example, at Bexley I interviewed students with autism, Asperger's syndrome, visual impairments, epilepsy, dyslexia, speech impairments and hearing impairments. At Cromwell, however, this diversity, in terms of impairment, was much more difficult to find, and most of the students I interviewed were either identified as having dyslexia or were wheelchair users.

19   Following publication of the Dearing Report in 1997, the government set a goal of 50 per cent of all British school-leavers going to university. Actions taken in response to this report, among other factors, led to a surge in post-16 student enrolment figures.

20   For example, many are named after Christian Saints.

largely autonomous institutions and whose physical location is less amorphous and more central to its identity.

Unlike Cromwell, which is still attempting to overcome its historically decentralised structure, the Central Administration at Bexley can, for example, make disability policies and expect to have compliance from all of the college's various sites. Conversely, because it is not as hierarchically structured as Cromwell, and has no history of elitism or exclusivity, voices from outside of the central administration do filter up to the top levels of institutional management. Individual members of staff and tutors can influence institution-wide cultures, whereas at Cromwell, the disability officer does not appear to have the same sort of influence. Indeed, the culture at Cromwell is so exclusive that one of the students remarked that Disability Resource Centre staff members are not taken seriously by tutors because they do not hold degrees from that institution. As she remarked,

> People in the Disability Resource Centre are not Cromwell graduates so the people who have to take note – lecturers who are famous in their fields – feel they are too smart to take note (Jacqueline, undergraduate student).

Whilst individuals at Bexley are encouraged to give feedback on how to make the institution more inclusive, at Cromwell, even the influence of the disability officer – whose post is very senior within the institutional hierarchy – is seen to be constrained by his or her own background and qualifications.

Another point of contrast is the availability of financial resources, an increasingly significant factor in institutional responses to laws like SENDA (Barnett, 1990). Bexley has very limited resources compared to Cromwell, and this affects everything from lifts not working to a lack of funds for making buildings more accessible. Bexley has the willingness to invest in issues of access but fewer resources; Cromwell is less willing to invest but has more resources with which to do so. Funding structures also impact these two institutions. At Bexley, students have to pay for their own transportation as the Learning Skills Council (formerly FEFCE) does not include financial allocations for transport to FE colleges, whereas at Cromwell these costs are subsidised by the university. This disparity between FE and HE funding regimes could indicate the relative importance of HE over FE (Mann, 2002).[21] At Bexley, students spend their own money, often from their monthly Disability Living Allowance, just to be transported to the college.[22]

In terms of their physical environments, infrastructure and equipment, Bexley has fewer financial resources than Cromwell and yet the college's efforts to be inclusive vis-à-vis physical accessibility are much more visible and obvious

---

21   Recent government initiatives are seeking to increase resources for the FE and Vocational education sectors, e.g. Tomlinson Report, February 2005. See www.dfes.gov.uk

22   The Disability Living Allowance is a social security benefit which may be claimed by persons under 65 who "have a long term health problem, mental or physical, that affect one's everyday activities" (Department for Work and Pensions, 2005).

throughout the campus. In any institution, there is a relationship between its ethos and physical structure. As (Hebdige, 1998:12–13) notes:

> Most modern institutions of education, despite the apparent neutrality of the materials from which they are constructed (red brick, white tile, etc.) carry within themselves implicit ideological assumptions which are literally structured into the architecture itself.

At Cromwell, the ancient buildings can convey the opposite of those on Bexley's campuses – they are old, forbidding, and many have narrow entrances and steep stairways. For example, wheelchair users at Cromwell cannot enter many buildings since the vast majority, over 70 per cent, do not have ramps or lifts and cannot easily be accessed as the buildings are listed with English Heritage (Estates Manager, in interview). Of course, there are ways of making listed buildings accessible, and English Heritage does have a programme specifically geared to increasing the accessibility of listed buildings, but such efforts require takes time, financial resources and the institutional commitment to make such changes.

As Lomas (2000) notes, buildings and other symbols of HEIs are important as they are linked to the identity and traditions associated with institutions. The physical spaces of an institution can reveal aspects of its unspoken values, ethos and culture. For example, the 'Old Joe' clock tower at Birmingham University is a point of pride for the institution and its continuing presence denotes a tribute to the past, and a linking of civic goals with educational ones (ibid.). In the case of Cromwell, as with other traditional institutions, there is a sense that the university conveys its elitism through its grand architecture, such as massive gothic and neo-classical stone college buildings, high chapel towers, vast courtyards, and ornate stained glass windows. While these can powerfully convey a sense of institutional history and its 'superiority', disabled students who cannot access these architectural environments may see them as cold, inaccessible and unwelcoming. Bexley, conversely, has an architectural style which is bland, concrete, Modern, particularly from the 1960s onwards, and functional. While it may not be as inspiring as Cromwell, there is a sense of flexibility and adaptability in the college buildings and surrounding environments.

The relationship between the physical spaces and structures of an institution, and its culture and power structures, is noted by many writers (Lomas, 2000; Radcliffe, 1999). As Radcliffe (1999:219) notes: "power and space/place are deeply intertwined" and they work in concert to form and influence social-geographical relations and processes. In institutional settings, power's multiple locations are spatialised and diffuse, and depend on institutional actors carrying out and reproducing power relations through daily actions and the (re)production of cultural norms and values (Bhabha, 1994; Foucault, 1979; Radcliffe, 1999). The disability officers of both Bexley and Cromwell expressed their frustration at the complex ways in which power relations at their institutions operate, which makes their jobs all the more challenging. In various ways, the interviewees expressed the view that they see their jobs – beyond making provisions for disabled students – as being about cultural change. For example, Bexley's disability officer said:

> I see it as my role to speak up and fight to change attitudes towards disability in this university. It is a tough battle, but if I don't speak up, who will? (Della, Disability Officer).

Given the importance of culture for students' experiences, this is certainly an immense task, and one which challenges the ways in which institutional actors view themselves. Additionally, the practices, ethos, attitudes and values of an institution itself, come from, reinforce, and are intermingled with, the space, place and buildings of the university.[23] As Hebdige (1988:11–12) notes, in the multitude of ways in which institutional cultures interact with the physical spaces of universities, "a whole range of decisions about what is and is not possible within education have been made, however unconsciously, before the content of individual courses is even decided."

To bring about changes in values and practices, Lomas (2000) proposes that strategies target both what he calls the 'culture' and 'structure' of HEIs. Additionally, Dopson and McNay (1996) posit that organisational culture is greatly affected by organisational structure and the distribution of power. Institutional structure, both physical and cultural, both need to be targeted by SENDA to create inclusion. Structure has a role in terms of how policies are actualised in practice and can affect the nature of provisions for disabled students. In considering the institutional cultures and structures of these HEIs, geographers would argue that "no matter how it is approached, "culture" is spatial" (Mitchell, 2000:63). Mitchell notes that culture "insinuates itself into our daily worlds as part of the spaces and spatial practices that define our lives" and that what he calls "new cultural theory" places great importance on space and sees culture as "constituted through space and *as* a space." (ibid.).

In this way, given the intertwining of institutional culture and space, the physical spaces and locations of institutions are significant as they affect the lived realities of disabled students. For example, disabled students at Bexley and Cromwell both face challenges in obtaining transportation to classes. For Bexley, the challenges come from its multiple locations which are dispersed over a large region. While Bexley is located in a city with a public transport system, the system is not fully accessible to people with mobility impairments, for example, and students face challenges in meeting their transportation costs. For Cromwell University, the issues around transport are entirely different because of its geography. As a more traditional campus, the buildings are located closer together than at Bexley, but the streets are narrower and less accessible to vehicles which can transport students in wheelchairs, for example. Additionally, because Cromwell University is located in a smaller city than Bexley, it has a less comprehensive public transportation

---

23   For more on notions of identity in relation to spaces of education, see Benjamin, 2002.

service for students to access. This indicates the difference that geography makes; SENDA's provisions have very little written about these geographical differences between institutions and how they might affect material outcomes for disabled students.

For example, taking on board Massey's (1994) theory that one's sense of place is mediated through various social, political and economic processes across a range of scales – and that it ought to taking into account local specificity, variability and contingency (without reifying difference), this law does not fully embrace this geographical notion that local institutional spaces matter and will lead to differentiated outcomes for disabled students.

I began this section with Barnett's (1990) typology for HEIs, which provided a means of differentiating conservative institutions from radical ones. However, a case in point for illustrating the incomplete analysis offered by categorising institutions in this manner comes from Barnett (1990:192) who repudiates his own typology of conservative-radical in higher education philosophies, saying that adhering to this typology would provide a "superficial analysis" and that it "would miss the subtle ways in which claims of liberality enter the inner life of institutions of HE." Using the example of access, Barnett notes that employing a liberal educational conception of developing the mind, conservative institutions opt for selective systems of entry based on formal demonstration of prior intellectual accomplishments, e.g. the use of A-level results to determine admission to Cromwell. In a 'whole person' model of education, which is less conservative, institutions such as Bexley may be open to taking extracurricular achievements into account, such as in sport, the arts, employment or the voluntary sector.

However, Barnett (1990:192) notes that in both cases "selection will tend to have a class bias".[24] Indeed, as hooks (2002) notes in her book *Where We Stand: Class Matters*, everyday interactions in institutions can reproduce structures of class hierarchy while simultaneously denying its existence. In the field of geography too, these issues of class have been debated, for example by Henderson and Sheppard (2006:68) who note that "in Marx's analysis, class location shapes interests and identities" and that material inequalities are related to spatial location.

Barnett's (1990) conservative-radical typology is limiting and institutions are more complex and ambiguous than this; they cannot be easily slotted into such categories. In the case of Cromwell and Bexley, they both employ liberal notions of access, albeit in very different ways. For Cromwell, its aims are based on the development of students' minds, intellects and the search for objective truths. At Bexley, liberal educational goals are focused on the development of the student as a whole person as well as growth in vocational skill development. This is linked to the creation of employees for the workforce, which can itself be seen as a liberal capitalist notion (Taylor et al., 2002). The next sections examine institutional

---

24   Barnett (1990) says that a more radical approach would be like that of the Open University, in which the imposition of admission criteria is greatly reduced.

contexts (5.4) and responses (5.5) to SENDA in light of broader liberal educational and legal philosophies.

(5.4) Discourses and Dilemmas: How HEIs View Disability and SENDA

> There are barriers in our work. Some colleges really do feel that the service they provide to their students is second to none. They had disabled students in the past, they didn't need any help then, and they don't see why they need help now (Martha, Disability Officer, Cromwell University).

This section examines how HEIs view disability and disabled people by unearthing and interrogating attitudes, values and practices towards disabled students and their inclusion within the two case study institutions. This is important because the values of HEIs influence all aspects of institutional life – from curriculum to governance, and the physical and social structures of campuses (Hebdige, 1988). HEIs in England, and indeed in other countries, have certain institutional philosophies and beliefs, which date back to their origins (Andrews, 2010; Bailey, 1977; Duryea, 2000; Kells, 1992). These go beyond pedagogy and have more to do with value systems of the institutions, what kind of students should attend, what kind of social activities they should engage in, and expectations about what students will go on to do, in terms of career choices, after their studies (Arnold, 2002).[25]

Institutional values are intricately linked to how HEIs engage with, and include or exclude, disabled students. Indeed, as Gourley (1999:84) notes, "the manner of engaging with diversity issues within universities tells you everything you need to know about the way they perceive their relation to society, and their ethos." At Cromwell University, the disability officer feels that the institution's ambivalent attitude towards disabled people is inter-woven with its liberal values and long-established traditions. For example, I had difficulty finding Cromwell's Disability Officer due to poor signage, and asked her why that was. She acknowledged the difficulty that many students have in finding her office, and felt that the lack of

---

25   For example, in her work on the career paths of students at elite institutions in the USA, Arnold (2002) asserts that students who attend Harvard, Yale or Princeton Universities, are expected to (1) be from among the top achievers in their high school classes, (2) actively engage in philosophical discussions with students and faculty outside of lectures, and (3) go on to prominent and successful careers as leaders in the fields of law, the corporate world, or public service. She also notes that beyond the regular teaching and learning which goes on in lectures, students are expected to connect meaningfully with at least one member of faculty, who can become a mentor for them, and a key factor in their academic and career success.

clear signs reflects the university's unspoken traditions of exclusion.[26] As she noted:

> I think you get the idea that you should know where you're going ... it's something that, I think when you come from outside, it particularly struck me, it's an established institution and it sees itself like that ... although on the other hand it's very liberal foundation sees itself as a very open place (Martha, Disability Officer).

This distinction being made between insiders and outsiders at Cromwell, i.e. that people in the institution ought to know where they are going, is a clear example of Marshall's differentiation (1990). This differentiating attitude draws boundaries between those who belong to the institution, and those who do not (hooks, 1994). Some human geographers contest the notion of boundaries, arguing that they reinforce exclusion; according to Massey (1993), boundaries "make distinctions between 'them' and 'us' and therefore contribute to a reactionary politics" (in Cresswell, 2004:73).

When the disability officer noted that Cromwell sees itself as a liberal institution, she equated this with openness to difference, and suggests that people in the institution see liberal values such as equality for all, freedom of speech, and academic freedom in a positive light. For example on the Equality and Diversity section of the University's website, it says,

> Cromwell University is committed in its pursuit of academic excellence to equality of opportunity and to a pro-active and inclusive approach to equality, which supports and encourages all under-represented groups, promotes an inclusive culture, and values diversity.

Indeed, as argued in Chapter 1, there are positive elements within liberal philosophies which can be beneficial to disabled people. At the same time, a more historically and epistemologically critical view of liberal values in HE would see conceptions of equality as being based on white, heterosexual, able-bodied men as the norm or the benchmark from which all others are measured (Duberman, 2002; Phillips, 2004; Razack, 1998, 1991; Young, 2001). The fact that there are members of this institution who on some level do not mind if the Disability Office is not well sign-posted and is difficult to find represents quite an exclusionary attitude. Indeed, despite romantic notions about academia consisting of communities of collegial

---

26   For example, a study conducted by the University of Edinburgh indicated that out of Cromwell University's 10,491 full-time undergraduate students in 2001–02, only 0.3 per cent were disabled (the study defined disabled students as those receiving the Disability Students' Allowance), which was lower than the UK average of 2.0 per cent. For full details see: www.planning.ed.ac.uk/PISG/HEFCE03PI/Table7.xls

scholars, "there have long been excluded as well as included groups in the politics of universities" (Becher, 1988 in Deem and Johnson, 2000). For example,

> We have policies on how to 'accommodate' students' needs ... but for example, about dyslexia ... there will be an uphill battle to change perceptions. There is still the view that it's the student who has to change, not the teaching and learning. No policy can change that (Barbara, Education Officer, Cromwell University).

This quote reveals the implications of liberal values in HEIs. Liberal values should not be automatically equated with equality, and there is no such thing as value-neutrality; the academic class has a vested interest in maintaining its privileges (Foucault, 1975; Razack, 1998; Taylor et al., 2002). As Barnett (1990:7) notes, "liberal HE has rested on the assumption that objective knowledge and truth are attainable ... independent of narrow interests." However, he critiques ideas of impartiality – as do Clark (2001), Blomley (1994), Young (1990), and Razack (1998) – and notes that HE has become "a pivotal instrument in the apparatus of the modern state" and that the academic community has become "a pseudo-class in its own right, exerting its own partial claims on the curriculum" (1990:7). Barnett (1990) also argues that HEIs strive towards the pursuit of truth, knowledge and liberal education, and that they value autonomy and rationality. Indeed, the concept of autonomy centres heavily on rationalist and individualist able-bodied assumptions about education (Benjamin, 2002). For example,

> Disabled people are expected to work hard, just like everybody else, to sort out their own issues ... it's a very individualistic culture here (Barbara, Education Officer, Cromwell University).

Certainly, if students are thought to be capable and autonomous individuals, they should be able to find their way around buildings and find the offices they need (Barnett, 1990). But this liberal idea of autonomy does not foster a sense of openness and inclusion, and does not help disabled students trying to access services. Similarly, Clayton (2008:259) notes that "whilst overt exclusionary language and behaviour is often hidden" the spatial exclusion of marginalised people is visceral to those who experience it and notes that "'difference' is understood through the cognitive, practical and situated geographies of the city in relation to a public culture of place" (2008:261).

The above quote from Cromwell's Education Officer also implies that the institution's liberal ideas and policies – whatever their limitations – are not matched in practice, and that the reality on the ground is actually exclusionary and contradicts these liberal policies of equality and inclusion. For example, a Senior Tutor of at Cromwell notes that,

> Even though we have equal opportunities policies across the university, we get uneven results in different departments. For example, the education department

and especially the medical school are worse than others. To them 'disabled' means 'not able to be a doctor' (Donald, Senior Tutor).

It is precisely this contradiction which so many educational theorists have criticised – that liberal educational policies do not necessarily counter exclusionary practices in institutions, and that perhaps they even allow institutions to mask practices of discrimination, exclusion and domination (Brooks, 1997; Foucault, 1979). Given these subtle exclusionary attitudes, the limitations of liberal policies, and the difficulty of enforcing them at Cromwell, how does Bexley compare? As previously mentioned, Bexley is quite a different institution from Cromwell, with values that favour inclusion and openness to difference and seems to be more in keeping with Barnett's notion of the education of the whole person.

At Bexley, the disability officer, for example, found the institution to be inclusive of equality issues, and their sensitivity towards her impairment during her interview for the position convinced her that it was quite a different institution to others she previously experienced. When I asked her about Bexley's policies on inclusion, she shared some of her own history with me, and noted how her own experiences of impairment were valued by the College. She said:

Disability officer (DO): I think it's my background. I mean I was studying for my doctorate in clinical psychology ... then I had my own sort of terrible breakdown ... and I joined a user group for people in psychiatric hospitals who were trying to make their voices heard. So that gave me a real sort of experiential, sort of, of what it's like to be on the receiving end ... And I thought I'd never get another job, with a mental health history. But also when I came here it was very ... when I came for interview there was a disabled man on the panel. So I immediately thought, yeah, they believe that, yeah that's great. I actually felt that I could say that in the interview. They said, why are you interested in this job? So it wasn't only about my skills for the job, it was also about my own personal experience.

FV: So you brought that with you to your position?

DO: Yeah, which is quite interesting really and quite useful. And quite interesting for me to try and translate that into a policy decision in terms of making things happen. So that I can have a sort of view that we should listen to students ... and I think the college and this faculty has a real commitment to that. It's not just lip service ... with our recruitment policies, for example. We proactively seek people who have had experiences of being disadvantaged, in the broadest sense. So, class, race, gender, whatever. So, that's really nice to be able to do that (Della, Disability Officer, City of Bexley College).

While exclusion does still occur at Bexley, its particular history, values and ethos of service to the community render it very different from Cromwell, and make it inclusive of disability in many ways. Indeed, this inclusion is more typical of the

New Universities and HE/FE colleges sector than of Russell Group institutions. For example, Deem and Johnson (2000:72) found that the hiring of senior management exclusively from those educated at Oxbridge and the public schools "has been slightly lessened, though mostly in the 1960s universities, the one-time colleges of advanced technology and the former polytechnic sector." While the inclusion of a disabled person on a job interview panel may seem like a small gesture, it provided a sense of comfort and inclusion to the disability officer and thus aided the institution to acquire a senior member of staff who is disabled and committed to making the institution more inclusive. Another member of staff at Bexley concurred with the view that the college works hard to be inclusive, and said:

> City of Bexley College does a very good job to be inclusive. It's not perfect; inclusion can never be perfect. I started working with people with disabilities 25 years ago because my brother has a disability. I've been in this job for 4 years, and it's nice to work in an institution that's so supportive of these issues (Estelle, Disability Support Coordinator, City of Bexley College).

Indeed, as many writers note, it is challenging for institutions to become more inclusive, but gestures such as reaching out to historically excluded groups through proactive hiring processes, can lead to broader inclusion over time (Gourley, 1999; hooks, 1994; Linton, 1998). The danger, of course, is that such gestures can be tokenistic, or not indicative of deeper and wider change in institutions. For example, Gourley (1999), a South African academic who was Vice-Chancellor of the Open University, does not think that small changes are enough to make institutions inclusive and believes in the complete transformation of institutions to make them accessible to all. Regarding her own role, Cromwell's disability officer feels her institution can be tokenistic. She notes,

> I think unless you have disability as part and parcel of other things that go on you have to look at what is it that really worries universities. Disability? As long as they've got somebody to do it and they're happy with it ... it doesn't really attack the core of it (Martha, Disability Officer, Cromwell University).

As she indicates, the idea that one person can change an institution's culture is disingenuous and could indicate a lack of institution-wide commitment to inclusion of disabled people (hooks, 1994; Linton, 1998). Based on my postal survey research in Chapter 4, many disability officers feel that they are the lone champions for disability in their institutions.[27] Theorists also note that equal opportunities policies are often seen as the responsibility of particular individuals rather than the institution as a whole (ibid.). Cromwell's Disability Officer certainly indicated that disability issues are seen as her responsibility and they are not 'part

---

27   For more results of the survey of disability officers, please see Chapter 5.

and parcel' of the fabric of institutional life. Conversely, Bexley's Disability Officer feels that her institution works to be inclusive as a whole, but notes that her supervisor, the Head of Faculty, had to struggle to build up accessibility services at Bexley. As she says,

> The Head of Faculty is a really strong character. She's like a politician. She built up the service from a back street portacabin operation to what it is now, which is a mighty empire, and it's good that she is in charge of it. That one person has been quite a driving force.

Such struggles of 'lone champions' working on issues of equality are well documented in HE literature (hooks, 1994). It appears that Cromwell's disability officer is set for a long struggle to build accessibility services up to the standard at which it exists at Bexley. This struggle may be due to many factors including personality issues, Cromwell's particular governance structure (e.g. decentralised colleges), institutional size and location, and its exclusive reputation and culture. While the Disability Officer at Bexley and her boss, the Head of Faculty, are able to influence their institution, Cromwell's Disability Officer finds that the responsibility for anything related to disability is assigned to her.[28] Equal opportunities initiatives at Cromwell do face challenges. For example, in 2001, Cromwell did a study that indicated it had a very 'macho' and exclusionary work culture. As a report on the study noted:

> Cromwell University has been criticised as having a 'tough macho culture' in a far reaching staff survey. The 'equality audit' among staff employed by the university found a high proportion of women, ethnic minorities and disabled people felt undervalued or excluded at some stage. Three quarters of staff said their employer needed to make progress on equal opportunities to live up to modern expectations (Cromwell Network, 2001:1).

Given the views of staff at Cromwell on the general lack of progress on equal opportunities, what are the specific attitudes encountered by disabled people? There is no doubt that historically, and even in the present, one can find antipathy for, and antagonism towards, disabled people in HE (DRC, 2005). At Cromwell, the Disability Officer acknowledged that exclusionary attitudes do exist, and could even affect her. For example, she said:

> If I go to them and say 'excuse me, you're discriminating against me and you're not observing my rights', I might get a negative kickback because they're doing

---

28   This is not how SENDA envisions inclusion in HEIs. According to the provisions of SENDA, each department, both academic and service-providing (e.g. halls of residence) must examine its services to students, and alter these, in anticipation of the needs of disabled students.

everything they can for disabled people. It's hard, but you just have to deal with it (Martha, Disability Officer, Cromwell University).

While the exclusion of disabled people is less common at Bexley than at Cromwell, it does occur. For example, the Disability Officer notes that,

> I think it's dominated by the faculties. So there are some extremely traditional faculties around ... and other faculties are more well known for being quite discriminatory so if they somebody would say dyslexia, they would think they shouldn't be doing a higher level course. They sort of associate that with being not so intelligent, which is dreadful (Della, Disability Officer, City of Bexley College).

These discriminatory attitudes are parlayed onto the spaces of HEIs, as evidenced by the challenges faced by disabled students in gaining physical access into and around buildings at these institutions. For example, at Cromwell, for years, wheelchair users who wanted to enter the Disability Resource Centre had to go in via a back entrance through the car park.[29] As Glendinning (1999) notes, a lot of writing on issues of access tends to "reduce the social oppression" of disability to problems of design in the built environment (in Gleeson, 1999:105). However, it is not simply a dichotomy between social oppression and a built environment problematic; both are deeply intertwined and the physical exclusion of disabled people in HE is a manifestation or reflection of those socially exclusionary attitudes.

While Bexley seems to exhibit more inclusive attitudes than Cromwell, issues of physical access sometimes 'slip through the cracks,' as one of their members of staff put it, for example, in their building works. In one of their brand new buildings, both lifts are often out of service, meaning that students in wheelchairs wishing to access classrooms (only the library is on the ground floor) must use service lifts ordinarily used for rubbish and freight. The building also features a lot of shiny glass, which can be confusing and dangerous for students with visual impairments. While in both of these cases, these buildings may or may not be legally acceptable in terms of SENDA,[30] they indicate a wider lack of consideration and inclusion of disabled people into the processes of building design, planning and construction (Imrie and Hall, 2001).

In addition to attitudes and values evident in subtle signs and symbols, and the blatant exclusion of disabled people in terms of physical access, some very revealing aspects of institutional life are evident in daily interactions and occurrences at HEIs (Holloway, 2001; Konur, 2000). For example, how are disability officers

---

29 This situation finally changed in the summer of 2003, when the Centre was moved to a more accessible and 'dignified' location. For more on the issue of dignity, please see Chapter 6.

30 Such specifics can only be determined when there is case law for comparable situations.

viewed within these institutions? At Cromwell, it has been subtly suggested to the Disability Officer that she maintain a low profile for her services so as not to attract more disabled students. As she says,

> You do have ... a Disability Coordinator, you do have a dyslexia service, but you keep quiet about it because if people get to know about it we'll attract more of those sorts of students, and do we want to really? And if you argue it, if you bring it out in the open, most people won't necessarily ... if you say 'is that what you're saying?' they'll draw back from it ... I see that as one of the things, since coming here, to change that. It's not that easy. Coming to change is not so easy (Martha, Disability Officer, Cromwell University).

The trickiest aspect of this, according to her, and others, is that much of these discriminatory attitudes are unspoken, so that people may not actually say such things, but rather hint at them or suggest them. It can be very difficult to challenge people on such ideas when they do not articulate them. The difficulty of unearthing discourses is detailed by Foucault (1979:49), as noted earlier, in his quote about how discourses are "practices that systematically form the objects of which they speak" and how they "conceal their own intervention." As Razack (1998) and Young (1990) note, subtle and covert exclusionary liberal discourses can be the most challenging to unearth and challenge.

In terms of legal requirements to counter discrimination, it may therefore be quite difficult to uncover these more subtle forms of discrimination with SENDA, which is primarily focussed on direct rather than indirect discrimination (Doyle, 2000).[31] It is very difficult to challenge or counteract such discriminatory statements or attitudes when they lie just beyond the realm of the overt. As Foucault (1977) notes, liberal systems of domination are perpetuated because they conceal their own exercises of power, and because they are portrayed as normal, everyday and mundane.

Another problem is that institutions sometimes use what Marshall (1990) terms rationalisation, e.g. senior managers at Cromwell could explain that they do not want to attract more disabled students because they have limited resources or wish to maintain standards. This does happen. For example,

> I still have to fight to persuade the university that costs for accessibility service should be brought into central funding streams ... their attitude is wrong and unlawful ... they don't understand the anticipatory duty (Donald, Senior Tutor, Cromwell University).

---

31   The Equality Act [2010] seeks to remedy this by adding indirect discrimination to the statutes; this takes the form of discrimination by association rather than an ADA-style provision for class action. For more detail on this new clause, see the Epilogue.

As Marshall (1990) notes, rationalisation is used to legitimate particular forms of merit and exclude others. As seen in this section, attitudes, values and practices towards disability and disabled people in HE can be discriminatory, exclusionary and sometimes opaque, making them difficult for disability officers to challenge. While my research at Bexley indicated more openness and willingness to engage with disabled students and to transform the institution in order to include them, many challenges still exist, for example, in fundamentally altering the ways in which building works are conducted, and getting academic staff to improve their attitudes. At Cromwell, these challenges are augmented by the institution's very traditional values and long history of elitism which is predicated on exclusion, and its status as a bastion of class privilege. As Cromwell's Disability Officer said:

> I do happen to think that Cromwell University actually for all its negative press ...
> I think actually it does try very hard. And the reason I think it does try very hard,
> and this is going to sound like a bit of a contradiction, is because of the traditions
> here, and their traditional values. Whether you like it or not, and I come from a
> rights perspective background, (in terms of) disability, you 'look after' disabled
> people. And it may be that that's not coming from a rights perspective, that there
> is a value judgement that's made that you 'look after' disabled people ... you
> haven't got a problem as long as you address the issues in their own language
> (Martha, Disability Officer, Cromwell University).

Employing notions of history and tradition, the institution makes efforts to be inclusive of disabled people, but will couch these efforts in their language of 'looking after' all students, including disabled students, simply because their students are the best and brightest, and should therefore be fully attended to. When I asked the disability officer, Martha, about whether students might find the language of being 'looked after' patronising, she said perhaps, but that at least their needs are being met, and they appreciate that more than the sensitivities the surround language. Given these contexts of attitudinal issues at Bexley and Cromwell, the next section examines the policies and procedures created by these two institutions in response to SENDA.

(5.5) Institutional Resistance and Strategies of Transformation

> Institutions are responding to SENDA rhetorically, but not in terms of any real
> action (Tim, Lecturer, City of Bexley College).

This section examines what institutions are actually doing in response to SENDA in the context of the attitudes and values discussed in the last section, and what forms of resistance such practices meet. There is a history of resistance to policies and procedures due to strong notions of academic freedom and territoriality (Bailey, 1977). As Deem and Johnson (2000:68) note, senior managers in HEIs face opposition to the introduction of change from "intransigent, territorial and inwardly focused academics." As in earlier sections of this chapter, it is important to locate from *whom* and *where* in institutions this resistance comes, and to consider the spatiality of power relations in HEIs.

In addition to disability officers, many other actors within institutions are creating policies and procedures to be more inclusive of disabled students, both in terms of access to buildings as well as curricula. As noted in Chapter 4, institutions in England and Wales have a range of responses, varying from taking no action, to making significant changes at their institutions. Many of the survey respondents noted the challenges they faced in getting their institutions to engage meaningfully with issues of disability and disabled students. Likewise, in examining Bexley and Cromwell, this section analyses institutional actions taken in response to SENDA, as well as the forms of resistance these actions meet. As in the previous section, this analysis reveals the importance of specificity and the difference that an institution's location, cultures and values can make.

The differences between Bexley and Cromwell, in terms of their approaches to implementing SENDA, are vast. For example, with regards to access to premises, because of its large campus and very large stock of listed buildings, Cromwell was able to secure a 'blanket grant' from HEFCE to increase accessibility 'across the board.' According to its Senior Architect, who leads disability issues as part of the Estates Management team, Cromwell received £700,000 for changes to premises, on the condition that the university conduct a disability audit for its entire building stock. In keeping with this requirement, Cromwell conducted an audit of its 160 buildings, and calculated that the cost of upgrading all premises in accordance with the British Standard 8300[32] would be £24 million, or £52 per metre squared.

The £700,000 grant therefore represents only 2.9 per cent of the total funding needed to make their university accessible. From where will Cromwell University, which has an operating deficit in the millions of pounds, obtain the other 97.1 per cent of funding it needs to make the rest of its buildings accessible, and what happens to students trying to access those buildings in the meantime? This shortfall does indeed create real problems of access, as evidenced by student testimonials in Chapter 6. One example noted by staff is:

> Some students come into our office really upset about something that happened to them at their college ... there are two or three colleges that don't want us to intervene on students' behalf. There is nothing we can do for them. There is

---

32   BS 8300 is considered an acceptable benchmark to which changes to premises should conform, according to the DDA Codes of Practice (DRC, 2005), though this by no means represents a fully idealised conception of access. The British Standards Institution is a design standards organisation, which facilitates the setting of standards, inspections and quality management. For more on the BSI, please see website: www.bsi-global.com

still the view that colleges 'own' their students[33] (Martha, Disability Officer, Cromwell University).

This is partly due to a lack of resources, the particular college system at Cromwell, and resistance to policies designed to include disabled students. As noted in the last section, lack of access it is not merely a built environment problematic (Glendinning, in Gleeson, 1999). Additionally, when students face problems with physical access at Cromwell they have challenges communicating these problems. As the Senior Architect noted:

> There is no formal communication or complaints procedure. There could be students struggling that I don't know about. I rely on networking to be told where there's an access problem (Glen, Senior Architect, Cromwell University).

A number of factors, such as institutional resources, an old building stock, the college system, and a lack of communication networks, can conspire to leave issues of access unresolved. At Bexley, with a much smaller and newer building stock, the issues of inadequate funding are similar while the approaches taken by the institution are different. Rules from the Learning Skills Council and HEFCE regarding funding create complications and limit the institution's ability to adequately fund its SENDA initiatives. For example, the Disability Support Coordinator at Bexley's Central Campus noted that,

> In order to receive funding, we have to conduct audits and provide proof. You can't buy equipment, you have to rent it; sometimes you spend more on rent than it would cost to buy, but those are the rules. We only receive funding for something if a student needs it, so we cannot meet SENDA's anticipatory duty.[34] Ironically, the reason you can't buy equipment, is because it's funded against a particular student and they will leave eventually. So we rent chairs and computer equipment, for example, at greater expense than what it would cost to buy, and we don't get to keep the supplies for other students (Estelle, Disability Support Coordinator, City of Bexley College).

---

33   Jacqueline, an undergraduate student at Cromwell, had to be carried up and down stairs for months, as no funding or other resources could be used by the Disability Resource Centre (in fact, she said the people at the DRC were "powerless"). The problem had to be addressed at the college level. For more detail, please see Chapter 76.

34   SENDA's duty to make reasonable adjustments is an anticipatory duty that means that institutions cannot just wait until a person with a specific disability applies for a specific course before making an adjustment. The institution should anticipate what may need to be done and make adjustments accordingly. This is an evolving duty in order that the base line for 'reasonable adjustments' is continually raised over time to ensure that institutions continue to improve their provision over time (HMSO, 2001).

Furthermore, despite conducting disability audits throughout Bexley's five campuses, certain key issues were forgotten. For example, at a new landmark building on the central college site, the college failed to consider how wheelchair users would access the pavement as there were no kerb cuts in front of the building.[35] The last weekend before the building was to open on a Monday, a temporary wooden ramp was installed in front of the building to allow wheelchair users entry through the pavement fronting onto the building. Additionally, in a brand new classroom which was purpose built on a different site, the college discovered that the metallic roof created a jarring noise when it rained, disturbing sound for students with hearing impairments during lectures. As one lecturer said,

> It seems so obvious that with a tin roof, you can't hear lectures if it's raining ... why didn't they think of this? (Tim, Lecturer, City of Bexley College).

Such examples illustrate the inadequacy of the particular audits conducted to consider all the possible ways in which disabled users interact with the physical environments of the institution (Imrie and Hall, 2001).[36] There is also the wider issue, evident at both Bexley and Cromwell, of how SENDA has been created without adequate provision for funding and other resources for institutions to carry out their visions of inclusion. Indeed, many authors write about how having goals for equality without adequate funding greatly constrains institutions' abilities to achieve these goals (Taylor et al., 2002). As Taylor et al. (2002:158) note, non-elite institutions like Bexley are "financially very hard pressed," and even institutions in the Russell Group carry large deficits. For example:

> Universities are under-funded; we need resources. It's an area of public investment that either has to come from general taxation or by charging a higher fee. I am the secretary-general of the Senior Tutor's Committee. The University is in favour of top-up fees; it's just that we regret the government's policy and way of going about it[37] (Donald, Senior Tutor, Cromwell University).

---

35   This oversight was explained to me in the following way: because the kerb is technically not part of the building, it was not included in the architectural designs.

36   Audits have become de rigueur in HEIs. For example, there are a number of bodies, including the QAA, Ofsted and Offa, which oversee the education sector and review and 'audit' the progress of individual institutions in relation to other, where the results are published in League Tables. These practices have been widely criticised as burdening staff with more administrative tasks and paperwork while failing to improve institutions. See Power, 1997.

37   What many HEIs resent about the top-up fees is the perceived administrative burden of now being monitored by Offa – the Office for Fair Access, which they say will not necessarily lead to inclusion, but rather a quota system. See Curtis, 2005.

Additionally, the constraints of managerialism, including extensive audit requirements, add to the feeling that SENDA is not specific enough, and does not pay adequate attention to the notion of inter- and intra-institutional difference and particularity.[38] For example, Cromwell's Senior Architect argues that in his view, adhering to SENDA itself does not guarantee inclusive, accessible buildings. He said:

> Our consultants on buildings ... follow legislation. They think they've complied and that's good enough. But it's not[39] (Glen, Senior Architect, Cromwell University).

He also argued that Cromwell's particular form of college governance, with sixteen autonomous colleges under the wider university remit, frustrates his efforts to create a more accessible campus. He noted that

> the colleges are self-administered, self-financed and inward-looking ... students are frustrated. They should come to see me, but they probably don't know I exist. The colleges have no money; in fact they hardly have enough for access audits (Glen, Senior Architect, Cromwell University).

The college system at Cromwell seems to be an example of Marshall's (1990) differentiation, whereby stratification within an institution leads to the dispersion of authority and accountability, so that large, centrally-mandated diktats become challenging to enforce in such differentiated institutional environments. This is also an example of his theory of institutionalisation, whereby certain procedures and power arrangements become static and crystallised in daily institutional life. Finally, in relation to Marshall's (1990) five processes for unravelling discourses in HE, Cromwell's colleges system indicates a lack of compliance, the fifth and final process he identifies.

As Schumaker and Carr (1997) note, if an organisation is so large that its members cannot communicate directly with each other then there are likely to be sub-cultures in which staff have attitudes and values which differ from those of senior managers. Deem and Johnson (2000) also concur that HEIs can contain many sub-cultures and multiple layers of communication and authority, taking in differences in geographical location, subject area, and complex structures and hierarchical relationships. Indeed, the same kind of dispersion that occurs at Cromwell also exists at Bexley. As Hebdige notes, hierarchical intra-institutional relations are reinforced through the arrangement of campuses: "the categorisation of knowledge into arts and sciences is reproduced in the faculty system which houses different disciplines in different buildings" 1988:12–13).

---

38   Funding rules are applied universally, and only provide limited amounts of money, regardless of how much the actual need may be.

39   Similar arguments were made about the DDA Part III and how compliance alone would not equal inclusion. For more on this, see Chapter 4.

An example of this is shared by Bexley's Disability Support Coordinator (DSC) at the central college campus who notes that each of the five campuses has its own DSC, and that their efforts are not coordinated centrally. Of her role, she said:

> DSC: My job is to coordinate anything to do with support at the central college campus ... including liasing with tutors ... outside agencies, IT companies, equipment, mental health etc.
>
> FV: Would it not be more effective to pool your resources together for certain projects?
>
> DSC: Yes, but that's not the way things work here (Estelle, Disability Support Coordinator, City of Bexley College).

She acknowledges that DSCs on different campuses could be duplicating their efforts, especially in dealing with outside agencies and companies. She notes that because Bexley is the product of several institutions merged into one, there are political sensitivities which force the institution to maintain a semi-autonomous campus system. As Barnett (1990:91) notes, "HE, in its knowledge offerings, will be found to reflect the disciplinary interests of the educational class" and this class can be further fragmented along political and geographical lines. Bexley's campuses are not as independent as the colleges at Cromwell, but there are still areas in which the campuses could work together more closely. For example, in implementing SENDA, the emphasis has been on site-specific strategies rather than on institution-wide action. This may be due to "the tensions that exist in attempting to balance institutional coherence with the need to sustain levels of academic autonomy and freedom of action" (Deem and Johnson, 2000:69). This dispersion can slow down the progress of SENDA strategies. As the DSC noted, because administration is spread over five campuses:

> There is not adequate administrative support ... sometimes it takes longer to relay information to someone. It's also quite intricate. (Estelle, City of Bexley College).

While SENDA's general aims to increase physical accessibility can be frustrated by institutional dispersion and college networks, challenges are also present in HEIs' efforts to make curricula more accessible. Curriculum itself is a contested notion – there is no such thing as a value-neutral or apolitical curriculum (hooks, 1994). As Taylor et al. (2002:155) note,

> Disinterestedness is not, however, possible, since the scholastic position is concerned to reproduce itself and its point of view and is, of course, always ideologically and socially constructed, and thus by definition partial.

Benjamin (2002) and Razack (1998) note that curriculum has tended to favour able-bodied people in its assumptions and perspectives. Taylor et al. (2002:159) also illustrate the importance of challenging the 'taken-for-granted' aspects of curriculum, when they write,

> The territory of higher education is in reality not based upon a consensual, let alone permanent, agreement on programmes, objectives, curriculum, approaches to research, priorities, epistemology and all the other features of the system. On the contrary, it is all highly contested, reflecting the contestation ideologically in the wider society. It does, therefore, matter profoundly what ends are served by the democratising processes of widening participation.

Theorists posit that academics feel threatened by increasing scrutiny of their work, particularly in teaching and learning as this is viewed as a 'sacred cow' (Barnett, 1990; Taylor et al., 2002). As noted in Chapter 4, SENDA is seen to increase administrative workloads of academics, added to a long list of regulatory responsibilities, including having to create reports for the QAA, RAE and internal research appraisals. In the particular case of Cromwell, lecturers were in favour of auditing teaching and learning but the leadership were not. Cromwell's Education Officer noted that the university's conservative leadership actively opposed changes suggested by teaching staff. She said:

> Lecturers asked 'why can't we have audits of teaching and learning?' Well, that was in the original plans in response to SENDA. The people on the ground are actually willing but the leadership people are not (Barbara, Education Officer, Cromwell University).

As at Cromwell, there have been varied responses to SENDA at Bexley. At Bexley, which has multiple campuses, part of the task of trying to change the culture is the challenge of trying to communicate with faculty and staff across the various locations. The fact that different teaching departments are geographically dispersed is also linked to a kind of cultural dispersion and some theorists even go so far as to describe academics in different disciplines as belonging to different "tribes" within their institutions (Deem and Johnson, 2000:74). Deem and Johnson (2000:74) also note that "the physical location of a department also affects the sense its members and head have of belonging to a wider community." Some departments see themselves as quite independent from the central administration, and run their departments within the framework of their own particular cultures. Bexley's disability officer commented,

> Some people don't even know that disability support exists ... it's quite incredible, they're in their own little kingdoms. The role of our faculty is supposed to be a cross-college, is supposed to be really accessible (Della, Disability Officer, City of Bexley College).

She gives examples of students within some of these departments who struggle for months, not knowing that a disability support unit exists, and is there to assist them, simply because their own faculty or department, away from the central campus, has not communicated this fact to them. Such situations seem similar to those at Cromwell and indicate the wider sense of isolation and lack of access to information and assistance which disabled students can face. Canagarajah (1999) notes that because of the sometimes intimidating atmosphere of higher education, HEIs should strive to create spaces that counter the isolation of marginalised students.[40] In the cases of both Bexley and Cromwell, members of staff feel that the efficacy of their efforts to be more inclusive depends on the ability of students to gain a mastery of the particular institutional cultures and languages of their HEIs. For example, the Education Officer at Cromwell said:

> I've learned the culture after three years. It's a complicated place ... I know how
> to speak the language (Barbara, Education Officer, Cromwell University).

Likewise, Cromwell's Senior Architect notes the importance of learning an institution's language in order to help transform it. He noted:

> I've been here for 16 years and seen 4 Vice-Chancellors go through the
> institution. You have to know the language. There are people who never learnt
> it, and they don't last (Glen, Senior Architect, Cromwell University).

In addition to learning the institutional language in order to transform HEIs, some key actors feel that the existence of SENDA, regardless of its legal substance, can be used as a lever, to 'goad' resistant staff into changing practices. For example, there are real fears of expensive and reputationally-damaging lawsuits in the HE sector and to some extent this may encourage HEIs to be more inclusive of disabled students in order to prevent being sued (Lewis, 2002; Palfreyman and Warner, 2002). This may not represent the ideal path to the inclusion of disabled students, but may lead to improved material outcomes for them nonetheless. As the DSC at Bexley noted:

> Even with the DDA, we are probably doing what we've always done, but we're
> able to do it easier now. Now I go to the faculty earlier, I get them to do more
> work in anticipation of students (Estelle, Disability Support Coordinator, City
> of Bexley College).

As the Senior Tutor of a College at Cromwell said:

40　This isolation and intimidation does occur according to student testimonies in Chapter 7.

Legislation has made a massive difference; it's assisted me in getting a cultural shift. The University Disability Committee went to the Vice-Chancellor and said, we'll get bad publicity in the press, and we're not delivering quality service. I couldn't have done that without the law (Donald, Senior Tutor, Cromwell University).

According to this testimony, despite the criticisms levelled at SENDA – that it does not include enough consideration of institutional specificity, and that the government has not allocated enough funding to accompany the legislation – it appears that the threat of being sued can motivate institutions to change. Such changes can be made if there are actors within these HEIs who understand their particular institutional dynamics and culture, and can use these dynamics to make changes. SENDA depends on the ability of individuals to take up the inclusion of disabled students within their institutions and to phrase their changes in a language that is amenable and politically acceptable. There is no legal mechanism within SENDA to force such changes de jure.[41] This can only occur after discrimination occurs, and the onus is on students to take up such cases. Moreover, institutional responses to SENDA are still mediated in and through particular institutional networks, and, given the history of exclusion at these institutions, the form and content of policy and practice at these HEIs is still shaped by ableist attitudes and values.

It appears that a greater proportion of disabled students are gaining access to institutions like Bexley, along with new universities and vocational streams at technical colleges, than to prestigious institutions such as Cromwell. As Taylor et al. (2002:158) note:

If the elite parts of the sector are not provided with incentives to participate in this agenda, then there is a strong probability that the hierarchical structure will be exacerbated.

The specific values, personalities and cultures of these two institutions mean that the same legislation, SENDA, is leading to very different outcomes in terms of policies, physical infrastructure, curricula and day to day operations. If the segregation of greater proportions of disabled students into less prestigious HEIs continues, then institutional specificity will be seen as an area for which SENDA failed to account, thus continuing a history of exclusion at England's top institutions, such as Cromwell.

---

41   The Equality Act [2010] includes a provision for public bodies to promote equality; this will require resources which do not appear to be accounted for. For further discussion of this act, see the Epilogue.

(5.6) Conclusions: Managerialism versus the Emancipatory Promise of HE

> Egalitarianism is naturally desirable when votes are to be gained, but it snuffs out the flame of inspiration and is the executioner of the first division (Vice-Chancellor of Cromwell University; Cromwell Students' Union website, 2003).

What does the above quote say about institutional culture, and the role of individuals in shaping that culture? How would one go about transforming an institution into an inclusive environment for disabled people, if senior leaders hold such views? Laws represent only one element of social and cultural change; to be effective they need to be implemented in enabling environments. If the VC is not committed to inclusion then SENDA alone cannot change the institution. Material outcomes for disabled students are still dependent on outdated able-bodied assumptions, perceptions, attitudes and values. This is not surprising given the way older institutions such as Cromwell operate in outmoded ways and in some cases are even proud of this traditionalism.

Institutions are increasingly concerned with centralisation, efficiency, and the development of corporate strategies and structures (Ball, 1990; Duryea, 2000; Henkel and Little, 1999). This, coupled with administrative burden, systematic performance measures, and increasing regulatory responsibilities, means that universities are severely challenged in keeping up with changes (Ball, 1990). As some analysts note though, universities have proven to have longevity, which suggests that they are effective at managing and adapting to change (Morley, 2004). At the same time, scholars in higher education note that universities can be very conservative organisations, intent on preserving their traditions and values (Komives and Woodard, 1996). Hence, they argue, the government has sought to intervene into the HE sector, with various reforms including the Education Reform Act [1988], the Further and Higher Education Act [1992], SENDA [2001] and most recently, the Equality Act [2010]. Regardless of the kinds of legislation created to change higher education, this chapter has shown that how those changes take place is dependent upon the contexts in which they are implemented, and by whom.

As Barnett (1990) notes, institutions which employ radical philosophies, like Bexley, are more likely to consider self-empowerment an important educational goal for students. The notion of empowerment, albeit linked to liberal ideas of autonomy, takes shape in very different ways at the two case study institutions. While Bexley works hard to be inclusive, its students' career and life options are limited by the fact of their attending that institution, as opposed to a much more prestigious one such as Cromwell. At Cromwell, a much smaller proportion of disabled students get in and succeed in that institution, but are promised access to greater life choices upon completion of their degrees. Problems still arise in terms of job prospects and discrimination upon entering employment (SENDA does not address or consider post-HE careers in this manner) but disabled students' chances of career success are greater with a degree from Cromwell rather than Bexley.

While this is also true for able-bodied students, the other barriers to disabled students entry into HE are greater, and therefore only a very small proportion of the already small proportion of disabled students entering HE will attend prestigious institutions. This indicates that the creators of SENDA either did not consider such specific institutional differences, or felt that this legislation could not address such issues of institutional difference and specificity. The result is that the inclusion of disabled students at Bexley and Cromwell has taken very different forms due to their institutional contexts and geographies. Delving further into the importance of geography and specificity, the next chapter examines issues of access to HE from the points of view of disabled students.

Chapter 6

Storytelling; The Voices of Disabled Students in Higher Education

(6.0) Inserting Voices into Discourses

In her book *Looking White People in the Eye: Gender, Race and Culture in Courtrooms and Classrooms*, Razack proposes an alternative to liberal narratives of law using a method called storytelling. As she notes,

> In the context of social change storytelling refers to an opposition to established knowledge, to Foucault's suppressed knowledge, to the experience of the world that's not admitted into dominant knowledge paradigms" (Razack, 1998:36).

Razack (1998) seeks to insert the voices of those who might be considered to be outside of the legal mainstream. In so doing, she seeks to destabilise categories of insider/outsider and lawmaker/legal subject, by uncovering what she calls narratives of storytelling in courtrooms and classrooms. Lawyers tell stories, as do teachers, as do legal witnesses and judges. Like Clark (2001) and Bellow and Minow (1996), Razack posits that storytelling is not a new technique; the stories which shape legal discourses are simply those which have succeeded in being heard.

Likewise, in their book *Law Stories*, challenging the methodology of lawyers' accounts of their experiences which excluded clients' voices, Bellow and Minow (1996:1) note that "much of what matters about law eludes most academic writings." Noting that academic writing in the field of law includes a range of accounts from the perspective of lawyers, judges, law students and professors, they argue that "despite this flowering of such narratives, stories of the actual experiences of clients and lawyers in concrete legal terms remain few and far between" (ibid.).

As a starting point for creating inclusion, Razack (1998) suggests the following: (i) uncovering the subjective narratives or stories which have been taken for granted in legal and educational settings, (ii) determining whose stories are considered legitimate and given material support by the state, institutions and courtrooms, and (iii) determining whose stories are left out, and then inserting *all* stories into educational and legal discourses. Likewise, Bellow and Minow, through their legal participant-narrative methodology, document stories which are inclusive of diverse perspectives, and "insights into how…those affected by law make their choices, understand their actions, and experience the frustrations and satisfactions they entail" (1996:1).

The use of a storytelling framework in this chapter portrays disabled students' experiences within the broader contexts of law, categories of disability and institutional practices. This portrayal is focused on *listening* to these marginalised voices as much as it is about *telling* their stories. Razack (1998), in examining courtroom evidence, demonstrates that women, aboriginals and disabled people have traditionally found that their stories are not heard or given due consideration when decisions that materially affect their lives are taken. She argues that the interactions between people in courtrooms and classrooms are based on liberal notions of equality, which, while egalitarian in name, are in fact exclusionary – an argument also explored in Chapter 1. For example, when in courtrooms, aboriginals must conduct themselves in ways that often do not take their cultural mores into account (Razack, 1998). In aboriginal culture it is considered disrespectful to look someone in the eye; yet aboriginals in court are told this is what they must do when taking an oath, for example (ibid.). Razack (1998) seeks to stretch the boundaries of the law to include different ways of human interaction which are not based on the dominant cultural norms upon which liberal legal systems are premised.

This chapter narrates stories about the experiences of disabled students in their own voices. In so doing, it seeks to illuminate some of the daily, lived experiences of disabled students in higher education in the UK. What little is known about disabled students in higher education suggests that their experiences have not been significantly documented or exposed (Caprez, 2002; Holloway, 2001; Konur, 2000). Students' testimonials in this chapter address issues relating to disability discrimination, rights discourses, conceptions of categorisation in education, and the geographies of disability as a lived experience.

The theoretical framework for this chapter comes from the concept of storytelling, which itself is built upon an edifice of critical legal, post-structuralist, and critical pedagogical theories (Benjamin, 2002; Fitzpatrick and Hunt, 1987; Foucault, 1979; hooks, 2003; Razack, 1998). Two main threads are interwoven with students' voices throughout the chapter. The first is the idea of using storytelling to disrupt conventional notions of subjectivity/objectivity. As Razack (1998:10) notes:

> These pedagogical directions make it clear that education for social change is not so much about new information as it is about disrupting the hegemonic ways of seeing through which subjects make themselves dominant.

The second main thread, which is interwoven with the empirical material throughout the chapter, is about how these stories and experiences are embedded in and informed by geographical notions of place and space (Blomley, 1994; Imrie, 2000). According to many geographers, space is conceived of as being more abstract than place. As Tuan notes,

> What begins as undifferentiated space becomes place as we get to know it better and endow it with value ... if we think of space as that which allows movement,

then place is pause; each pause in movement makes it possible for location to be transformed into place (1977:6).

Using Tuan's notion of space being defined in relation to movement, mobility is certainly a key ability for all humans (Urry, 2007). How mobility works for disabled students is one of the focuses of this chapter; further taking on Tuan's concept, when students pause from their motion through space, they will be in institutional places. Their reflections and perceptions may tell us something about HEIs as places.

(6.1) Traditional Discourses: Why we should Care about the Voices of Disabled Students

As mentioned in (6.0), two main threads present throughout this chapter are the use of storytelling to highlight disabled students' experiences in higher education, and how these stories are linked to larger issues of disability discrimination, rights, access and geographical notions of place space. This section engages with these critical debates so that they are clear in light of the empirical findings presented later. Moreover, it is important to illustrate why the dominant discourses need to be expanded to include disabled students' stories. Foucault (1979:49) notes that discourses are

> practices that systematically form the objects of which they speak. Discourses are not about objects, they do not identify objects, they constitute them and in the practice of doing so conceal their own intervention.

Traditional discourses of disability in relation to higher education have included the following: complete silence on the matter, exclusion, denials of the existence of disabled people in higher education, ableist and assimilationist attitudes, and patronising heroic portrayals of disabled people in higher education (Clare, 1999). For example, Professor Stephen Hawking of Cambridge University is thought to be a genius. However, much of the biographical literature about him mentions his disability and exclaims how remarkable his accomplishments are, all the more so because he is disabled (Kelly, 2003).

Higher education in the UK and elsewhere has traditionally been a bastion of privilege and power of the few over the many (Bailey, 1997; Bourdieu, 1998; Savage et al., 1992). This is particularly critical as education is seen as a key to succeeding in Western societies; the ability to participate in social life and to access economic goods have been shown to directly correlate to levels of education (Bourdieu, 1998). HEIs also contribute to the shaping of societal discourses in areas which have real impacts on disabled people, such as law, human rights, identity politics and other features of liberal democracy (Swain et al., 2004). If disabled people cannot fully and equally access higher education, they cannot move forward economically. For example, 35 per cent of disabled people were living on an income which was half the basic average income or less in the UK in 1999 (DRC, 2005). More recently, a study by the Children's Society indicated that 40 per cent of disabled children in the UK, a total of 320,000 individuals, live in poverty.[1] Without higher education, disabled people also find it difficult to engage in the socio-political debates which directly impact their lives and livelihoods, and to gain positions of power.

Ironically, for all its critical pedagogical tools, academia has been remarkably uncritical of the ways in which it has traditionally conceptualised disability, both theoretically, and in the day-to-day provision of HE to students. Perhaps this is not surprising given that

> it has been said that the teacher today is teaching what has been learned at least two generations ago from textbooks that were written several decades before (Dominelli, 1997, viii).

This view is shared by many educational theorists who note that the ways that HE conceptualises 'minorities' such as women, people of colour, gays and lesbians, and, disabled people – and the ways it does or does not seek to serve and include those populations – hearken back to traditions which mask power relations of domination, subordination and exclusion (Razack, 1998; hooks, 2003). Moreover, it is not merely inaccessible buildings or curricula which exclude disabled students from HE, but the actions of those in positions of power which can work to maintain the status quo and exclusion (Konur, 2000). Applying Tuan's (1977) theory about space and place to this context, when disabled students cannot access the spaces of HEIs, it can make their perceptions of these institutions as places inhospitable or exclusionary. This is in turn can affect their psychological and emotional well being, as well as their educational achievement. This fits with broader debates and discourses about liberal legal conceptions of the human subject; notions of cultural inclusion, and how 'rights' discourses affect people 'on the ground' (Young, 1990).

(6.2) Windows into Disabled Students' Lives

The findings from my interviews with disabled students are placed into two themes: (a) Category as experience; experience as category, and (b) Navigating the disabling attitudes, values and practices of higher education. The first theme is an exposition of the ways in which disabled students are labelled and categorised, and how they experience this categorization.[2] This is important as it brings to light the attitudes and assumptions students encounter from mainstream society and shows

---

1   www.childrenssociety.org.uk, 7 October 2011.

2   While many of these experiences are shared by disabled people in general, this chapter focuses on disabled students in higher education.

how disabling the language and practices of disability can be. This also illustrates how disabled people are not simply reducible to categories but rather, experience the world in rich and complex ways, with multiple and fluid identities, and with real differences between and among them (Clare, 1999).

The second theme, building on Chapter 5's examination of institutional practices around disability, highlights how disabled students navigate and negotiate their way around HEIs as sites of disabling attitudes, values and practices. These experiences demonstrate how the categorisation of disabled people is manifested in disabled students' daily encounters with tutors, administrative staff, other students, and the built environments of HEIs.

(a) Category as Experience; Experience as Category

> My department, they said, 'Oh yeah, we had someone who was dyslexic last
> year. No problem, we know.' Like so, they're pleased I was not a new animal
> for them, they had classified it already (Albert, a student at Byron University).

In researching disabled students' experiences, their constant battles against being categorised, labelled and judged are recurring themes. As discussed in earlier chapters on how the DDA employs medicalised definitions of disability, there appears to be a tension in liberal legalism between the need to treat all citizens equally and the ways in which the legal system identifies, differentiates and labels disabled people according to their needs or impairments. As disability studies scholars note, there has been a history of using elaborate systems of labelling and classification to exclude disabled people even if done in the guise of trying to meet these individuals' needs (Barnes, 1991; Gooding, 1994; Oliver, 1990). Authors in various fields, including geographers, feminist and anti-racist theorists, and critical pedagogists, have criticised liberal notions about, and methods of, categorisation (Benjamin, 2002; Freire, 1970; Gleeson, 1998; hooks, 2003; Knopp, 1998; Razack, 1998). For example, Knopp (1998:152), who examines sexuality and geography, argues that:

> the construction and essentialisation, through social practices, of variable human
> differences and experiences is a fundamental mechanism of social control.
> Control over various aspects of the production, reproduction and distribution
> of values (all broadly defined to include cultural as well as material products) is
> most easily exercised when people can be categorised and forced to conform as
> members of supposedly homogeneous groups.

The field of education has particular responsibility for what Knopp (1998) identifies as 'value production' and creates particular ways and experiences of knowing and being (Benjamin, 2002; hooks, 2002). These ways and experiences can be limiting for disabled people, and can also communicate ideas about the supposed limitations and deficits of disabled people in comparison to able-bodied students in education (Race, 2002). Being classified as disabled, and having a specific type of impairment closely intertwined with one's identity can be disabling, de-humanising and can limit disabled people's possibilities (ibid.). As one student said:

> We are definitely not a meritocratic society…I applied to all these gap year kind of things, but when people find out about my dyslexia and epilepsy…it doesn't exactly open doors for me (George, a student at City of Bexley College).

Several authors indicate that the educational experiences of disabled people in the UK have historically been exclusionary (Race, 2002 and Holloway, 2001). As Whittaker and Kenworthy (2002) note:

> The history of segregation, the practice of labelling and the denial of rights of learners have together stifled a dynamic shift towards inclusive education (in Race, 2002:68).

Successive governments from the 1950s to the present time have enacted various laws and charters to end the segregation of disabled people from mainstream education to little avail (ibid.). For example, the Education Act (1981), brought about in response to the recommendations of the Warnock Report (1978), was seen as a law which would finally end discrimination and segregation of disabled people in schools (Race, 2002). However, as the following brief history illustrates, this did not happen:

> The 1981 Education Act not only promoted greater integration but introduced the 'Statement of Special Educational Needs (SEN).' This placed a formal obligation on Local Educational Authorities (LEAs) to provide a written account of the child's identified need, the provision they would make to meet that need and the school placement they would offer. However, for many families 'statementing' has proved to be a painful and laborious process in which their child is categorised and labelled. Local authorities then used the statement to legitimise the rejection and exclusion of the child from their local neighbourhood schools. Thus, for these families the statement has not opened doors to new opportunities for their children but rather has served notice that the mainstream school doors remain firmly closed (Whittaker and Kenworthy, in Race, 2002:69).

This history illustrates how a system of classification and codification – in this case through the use of SEN statements – has material consequences for disabled people in education. Due to impairment, many of the students I interviewed do have additional requirements in order to facilitate their educational participation, and they do not oppose the idea that these requirements should be met. What they do criticise is how special needs have historically been conceived and categorised, and how this categorisation has been used to limit their educational opportunities, by placing pupils into special schools which do not necessarily meet their needs. For example, in my interview with Eric, I asked:

> FV: What do you think about the 'special needs' philosophy?

> Eric: I'm not actually too sure about that. I mean, technically, the human race's knowledge about autism and that sort of thing is rather limited in some ways (Eric, a student at City of Bexley College).

As Eric pointed out, knowledge about disability and impairment is limited and therefore SEN statements cannot be seen as conclusive evidence about children's abilities. Indeed, as Eric and others show, just by being in HE they may have surpassed expectations placed on them by LEAs. Having made it this far in an educational system that did not meet his needs is an accomplishment. Beyond the school's lack of understanding of autism, and therefore not necessarily meeting his educational needs, Eric had to survive in a difficult educational environment. The experience of being a student in a special needs school for Eric and others was horrible:

> FV: How was it at school?

> Eric: At school it was rather, well, how shall I put this? Violent and colourful (Eric, a student at City of Bexley College).

The experiences of being categorised, being placed into segregated 'special needs' schools and then finding out that these schools do not necessarily meet one's educational needs can be quite devastating. In Eric's case, it was also violent – something no student anywhere should have to experience. In his engagement with Young's (1990) "Five faces of oppression" – exploitation, marginalisation, powerlessness, cultural imperialism and violence – Gleeson (1998) points out that various marginalised social groups, including disabled people, experience these forms of oppression in different ways. He argues (1998:90) that while disabled people do suffer economic marginalisation, "unlike women or gays, disabled people are probably not subjected to systematic violence in capitalist societies." While it may not be systematic, all of the disabled students that I interviewed had been targeted by violence and other forms of abuse including verbal slurs and epithets.[3] As Young (1990) illustrates, the violence experienced by disabled people and other oppressed groups is shaped and informed by notions of the category. This oppression is experienced by disabled people not only as an overtly recognisable form of domination, but also in a more subtle manner in their day-to-day existence. Young (1990: 124) writes:

---

3   By Gandhi's definition, poverty is the worst form of violence, and disabled people are statistically poorer than able-bodied people (Brent and Lane, 1972; DRC, 2005).

> The objectification and overt domination of despised bodies…has receded in our time, and a discursive commitment to equality for all has emerged. Racism, sexism, homophobia, ageism, and ableism, I argue, have not disappeared with that commitment, but have gone underground, dwelling in everyday habits and cultural meanings of which people are for the most part unaware.

According to Young (1990), the experience of violence is part of the category of disability, as ableism has been suffused into everyday actions of which people are not always aware. Similarly, throughout their book *New Geographies of Race and Racism*, Dwyer and Bressey (2008) indicate how the category of 'race' can be spatially intertwined with hostility and violence. Among the interviewees, Eric and Chris certainly experienced physical violence in school, while all the others – Roger, Hamish, Jacqueline, George, Theodore, Peter, William and Frederick – experienced verbal and emotional abuse. As Chris, an undergraduate student in London who attended a mainstream school in the North of England, said:

> There were many bullies and I was involved in many fights in school…my educational experience was incredibly uncontrolled…I was always running from fights, and if I wasn't able to run, I fought. Being in school was a big problem (Chris, a student at Byron University)

Another interviewee noted:

> Sometimes when people don't understand what I'm saying they have a go at me cause they can't understand. They say something cheeky, or snigger and walk off with their mates (William, a student at City of Bexley College).

Beyond my interviewees' experiences, bullying and social exclusion of disabled people in education is well documented (Benjamin, 2002; Race, 2002). According to one survey, 38 per cent of students said they were bullied in school (DRC, 2004). This illustrates how the category of special needs has material consequences for disabled students – stigmatisation, and the expression of ableist attitudes in the form of school violence and bullying. As Clare (1999:68–9), a poet, essayist and disability activist, shares from her own childhood experiences,

> *Retard.* I learned early that words can bruise a body. I have been called *retard* too many times, that word sliding off the tongues of doctors, classmates, neighbours, teachers, well-meaning strangers on the street. In the years before my speech became understandable, I was universally assumed to be "mentally retarded." When I started school, the teachers wanted me in the "special education" program … I was *retard, monkey, defect*, on the playground, in the streets, those words hurled at my body, accompanied by rocks and rubber erasers … words bruise a body more easily than rocks and rubber erasers.

The language and categorisation of disability, then, has material impacts, as it shapes attitudes, ways of knowing and understanding, and actions. While Eric criticised the fact that his special needs school did not in fact meet his needs, and Chris pointed to the violence he experienced because he was identified as having special needs, Theodore found the term special needs itself to be problematic. He said:

> I was originally going to Institution X. However, I wasn't happy with the attitudes of staff, and their use of the term 'special needs.' Last year in September I was due to start there, and was talking to a member of staff about transport to get there...they kept talking about my 'special needs'...I did not like the attitude...I decided to leave it and have a look at what other sites (within the same institution) there were (Theodore, a student at City of Bexley College).

Theodore and the other students grew up in an era of special needs and their educational experiences were marked by horrible prejudice and battles to overcome this categorisation, and, in some ways, to achieve more than was expected of them. George, a student at City of Bexley College who aspires to be a politician, like his father, also dislikes the term special needs and the practices which accompany it. For example, he said: "I grew up in an era of Mencap," a charity started in the 1950s which spread patronising notions that people with learning impairments were mentally incapable and which classified people according to their purportedly measurable 'mental capacity' (Race, 2002).[4] George protested to his school authorities and asked them, "Why do you want to put an intelligent dyslexic person in a classroom with a person who is mentally 6 years old?" He talked further about his experiences, saying,

> I was placed with other people who had 'difficulties.' That's what they called us...can you believe it? It was assumed and expected that we would just get on well because we all had 'difficulties,' even though some of the pupils couldn't even communicate with each other. I was the best student in my class; others were doing far worse than I was (George, a student at City of Bexley College).

He felt he was wrongly placed with other students who were not at the same level of learning as himself, simply because they were all categorised as disabled. Again, the categorisation of disability and the language of special needs had material consequences – in this case segregation and institutionalisation for George and Theodore. The differences between Theodore and George are significant – George is mobile, articulate, confident, and has a learning impairment. Theodore uses a highly automated wheelchair, often needs a speech interpreter in order to be understood, and is shy and introverted. Yet they were both placed into special

---

4　For example, Clare (1999:105) describes the English charity Mencap as "posing as saviours of disability."

needs institutions that were socially isolating and away from their families. As George said:

> I was born before dyslexia was discovered. In primary school I was identified as having 'special needs.' I grew up in Wiltshire. My parents moved to Gibraltar when I was 4 years old. At 14 I was sent from Gibraltar to a special dyslexic boarding school in the Southwest of England. My parents were living it up in Gibraltar. They had a private beach, while I was in gloomy dismal England. I despised it intensely. My boarding school was 10 miles away from any town. I went from eating freshly cooked fish from local beaches to school canteen food of the 80s times two. This was what you had to eat day and night. I disliked it intensely. The only other option was a Gibraltarian English school which had no special needs unit (George, a student at City of Bexley College).

Both George's and Theodore's parents were acting on the advice of local authorities, and assumed that they knew best in keeping these children in special needs institutions. Indeed, many parents are pressured to comply with the 'diagnoses' found in SEN statements (Race, 2002). George's educational options were limited not only by this categorisation but also by his family's living arrangements. While he would have strongly preferred to continue his education in Gibraltar, he had no choice but to attend boarding school in England, as his purported special needs could not be attended to in Gibraltar. He was very unhappy in England and this distance caused friction in his relationship with his parents. Likewise, Race (2002) points out that there is a history of children being sent away to school because it was thought to be in their best educational interests. These students' experiences illustrate that however well intentioned, categorisation of disabled students and special needs schools do not always work to meet students' actual needs. As Albert said:

> I learned to read and write a bit later than anyone else. I was really bad at dictations and it caused me trauma. In high school and as an undergraduate I received 'accommodations' such as extra time for tests and having my spelling mistakes ignored. However, that's not always the solution to the problem (extra time). There are other aspects. I tend to read it kind of differently, for example I don't really realise that commas and dots are there. Sometimes questions are phrased in a way that I could interpret it differently. A useful accommodation would be to give a few options, or to be assessed differently. However, the only thing I always get is extra time (Albert, a student at Byron University).

Albert's story shows that a lack of understanding of different impairments, or simplistic solutions such as providing extra time, can work to the student's detriment. Dyslexia is not something that has one blanket accommodation[5] or solution; it varies for each individual. There are indeed differences and nuances, and each person experiences impairment differently.[6] Another student at the same institution, Chris, illustrated that the accommodation given to him was not only insufficient to meet his needs, but was actually counter-productive, frustrating, and in his words, 'useless.' He said:

> Providing extra time is useless for me as my concentration goes after 2 hours… In my last year of middle school to my first year of high school, I went from the bottom of my class to the top sets. The ability to communicate verbally became important to my teachers. When I was 13 and 14 my reading and writing were abysmal, I had no concentration. It was so frustrating – everybody else seemed to be able to do so much more in less time. I could see there was more and I couldn't get it – I had to fight for it. I have to take everything to the fullest extent – otherwise I can't do it (Chris, a student at Byron University).

Chris noted that as he progressed in school, the education system encouraged pupils to develop their verbal communication skills, which he excelled at, rather than written skills, with which he struggled. Had it not been for this later emphasis on verbal skills,

> I wouldn't be here today…they weeded out a lot of bright students simply because of reading and writing problems. The LEA felt that pulling me out of class for extra tuition was the answer to everything. I struggled and struggled in school, I didn't realise until much later that I actually have the intelligence – they just didn't know how to teach me (Chris, a student at Byron University).

This is an example of an education system that does not always work for students; it forces students to conform to the system, with very narrow conceptions of how different students can excel at learning and acquiring knowledge (Benjamin, 2002; Holloway, 2001; hooks, 2003). Disabled students like Chris and Albert, who are now at one of the UK's top universities, would benefit from an education system that values the skill-sets and knowledge they possess. Chris has to work so much harder just to keep up with his classmates. Likewise, Albert, an overseas student doing a postgraduate degree in London, found that the education system did not always help alleviate his reading impairments.

These stories highlight another issue within broader notions of categorisation of disabled people in education: expectation and achievement. Many authors

---

5   The term 'accommodation' is used to mean adjustments made to higher education procedures which take students' impairments into account. It is a widely used term in HEIs in the UK, USA and Australia.

6   As dyslexia is the impairment experiences by the majority of disabled students in HEIs, institutions will have to develop more nuanced understandings of it.

point to a history of the association of impairment with inability and lack of intelligence or intellectual capability (e.g. Mencap, as discussed earlier) and the use of categories to label and control various populations (Foucault, 1975; Razack, 1998). Employing a Foucauldian critique of management in education, the actions and decisions of LEAs regarding disabled students seem to indicate a belief that impairment can be mastered, measured and understood rationally and scientifically (Ball, 1990). According to these categories, disabled students were considered mentally and intellectually the 'lowest' in the educational hierarchy in terms of their abilities and what they could hope to achieve (Benjamin, 2002). Students like Chris have had to overcome low expectations of them by the educational authorities and struggled to stay in school, let alone getting into HE. As Chris said:

> I can't believe I got from that (school) to this place (university)... I could see there was more and I couldn't get it – I had to fight for it. If I'm even going to keep up, I've got to work damn hard, harder than everybody else. My mates spend 1 hour a day; I spend 4 hours (Chris, a student at Byron University).

The frustration in Chris's voice is evident. He is succeeding in university but the system is designed for able-bodied students, and therefore, not only does he have to work harder just to achieve the same results, but the whole culture of expectation and achievement for him and other disabled students is very different. Benjamin (2002) notes that for students who have been identified as having special educational needs, understandings of success and achievement are produced differently than for able-bodied students. She notes that the Department for Education and Skills (DfES), LEAs, and schools have different expectations of 'special needs' students, based on their understandings of 'disability,' 'identity' and 'difference' (ibid.).

Facing such discrimination and self-doubt, how did these students make it into HE? What spurred them on to enter university? For many of them, it was the influence of parents, teachers, peers and other supportive individuals. This calls into question liberal rationalist paradigms of human life on which SENDA is based, which paint a bleak picture of society as comprising independent, self-serving individuals (Altman, 1990). While there has been significant criticism of the Conservative Party's notion of a Big Society, it is possible that social functions can nurture individuals in ways which the state cannot.[7] Each student had very different experiences of support, encouragement and other ways of 'breaking out of their categories,' overcoming barriers to their entering higher education, with variations depending on their location, impairment, age, gender, 'race,' and even their personalities. For example, Chris's father encouraged him to go to university.

---

7   Of course, those who contest the notion of a Big Society are not necessarily philosophically opposed to the idea but critique this idea as it has been introduced in concert with, or perhaps as window-dressing for government spending cuts.

Chris said:

> My Dad told me, 'if you want to do something with your life, you have to go to university.' Both my parents did diplomas in Durham. They valued education (Chris, a student at Byron University).

Certainly, in terms of social and economic advancement, education is indeed valuable (Arnold, 2002; hooks, 2002; Razack, 1998). The majority of disabled people live below the poverty line, and education is seen as a way out of economic deprivation (DRC, 2005). Like Chris's father, Roger's father wanted a secure future for his son. Knowing that Roger would face struggles even getting to university interviews in a wheelchair, he wrote to all the universities to which Roger applied, to prepare them for 'accommodating' his needs and to ensure he could actually get into the buildings in which his interviews would take place. About his father, Roger said:

> His attitude is to get on with it, live life 'normally'…at 16 years old I started wheelchair use when I was in 6th form at school. I could walk to an extent. It was horrible. These days they would be utterly rapped over the knuckles for treating a student like that. I had to go up and down the main stairs several times a day literally crawling. I was knackered at the end of each day. My lessons should have been moved downstairs…my Dad didn't want these things to hold me back (Roger, a student at Cromwell University).

Roger has a hatred of the category of disabled as well as notions of normality and the idea that this might exclude him, coupled with a yearning to blend in, assimilate, and be part of the mainstream. As he said: his approach is: "I want to do what everybody else does, I want to be integrated into the mainstream." After studying law at one of the UK's most recognised universities, he went on to work as a solicitor at one of London's largest global law firms. Indeed, many disabled students experience conflicts around ideas of normality. Jacqueline and her sister both abhorred the elitism of Cromwell University but in the end decided that the potential for personal advancement gained by attending Cromwell would outweigh the negative experiences of being there. Both of her parents are academics and she said in some ways this provided her comfort and the knowledge that universities can have liberal[8] or progressive elements and also that in order for her to counter discrimination she would need effective tools – such as the ability to articulate and argue her position – which would come with a university education.

Like Roger, Jacqueline faced an internal struggle to put herself through negative experiences of education in order to gain more tools for success. This success is seen as partly a desire to assimilate and partly a desire to achieve more

---

8   The term liberal is used in contradistinction to understandings of what it means in legal contexts.

than is expected of disabled people. Foucault (1977) notes that in introducing certain norms and values in the nineteenth century, state and church institutions were successfully able to get citizens to adopt those values and regulate and police themselves, for example in the realm of work, economics and the imperative that poor people earn their keep (Foucault, 1977). Some disabled people also struggle to be normal, even to the point of sacrificing their own desires, just so they will fit in with mainstream society (Clare, 1999).

In addition to the low expectations placed on these students, the attrition rate for students categorised as disabled in the UK is very high. In fact, only a small minority of disabled people actually make it into HE – less than 5 per cent by one measure (DRC, 2004). This is well below the target set by the previous Labour government that 50 per cent of all students should enter HE. In England and Wales, the attitudes and practices accompanying the category of special needs have demonstrably led to lower expectations of and higher rates of attrition for, disabled students (DRC, 2004). Chris was the only one in his group of special needs friends to stay in school beyond the age of 12. As he said:

> I was diagnosed with dyslexia around 7 or 8 years of age. I was taken out of class once a week to a non-conventional woman who taught me and built a short-term memory for me. I went there until I was 12. In high school I was involved in a lot of fighting. I didn't have the work ethic. When I was 12 or 13, there were four or five others who were dyslexic. All of us went 'fuck it.' The other 4 or 5 went out of education, they had jobs (Chris, a student at Byron University).

Students like Chris, Albert, George, Jacqueline, Roger and Theodore are the exceptions to the rule.[9] Staying in education for disabled people requires a tenacity to deal with all forms of discrimination – from subtle to blatant. For example, one survey found that 25 per cent of students said they were discriminated against in school, and 20 per cent said they had been discouraged from taking GCSEs (DRC, 2004). The consistently low expectations of them in the education system, coupled with various forms of discrimination, can affect these students' own perception of their abilities and cause them to self-select out of achieving more. For example, Jacqueline, who completed her first year of Philosophy at one of the UK's top institutions, had a very positive experience in sixth form college. However, this was not enough to undo years of self-doubt about her own abilities due to discrimination she had faced. As she said:

> I'm 21 years old and I've just finished my first year at university. I left school early, and almost didn't come back. I went to private school. It was a specialist music school, one of only 4 in the country. I played cello until I was 17. Eighty

---

9   While it would have been interesting to interview students who did not make it into HE, the remit of this study did not allow such a study, due to limitations on time and resources.

per cent of students from there go to music college but I didn't (Jacqueline, a student at Cromwell University).

Jacqueline's particular circumstances, such as her musical talent, her supportive school environment, and her positive attitude, all helped her get into one of the most competitive universities in England. However, compromises were made along the way. For example, one of the reasons she did not pursue a career in music was because she knew there would be stigma for a wheelchair user playing the cello. In classical music recitals it is quite rare to see a wheelchair user as part of the ensemble, and Jacqueline feared being discriminated against in an already difficult and competitive field. She also had battles in school before she got to sixth form college but said that she worked hard to maintain good relations with the school as it was in her best interest to do so even if they didn't always treat her with respect. These are some of the subtle daily humiliations disabled people, and other marginalised people, can face, for example, putting a brave face on inequality and discrimination to further their long-term career interests (Matsuda; 1993; Morris, 1991; Razack, 1998).

Living and experiencing life as a category can be limiting and disabling. For the students I interviewed, having been categorised since childhood was a foundational and formative experience, and they came to know and experience the world in and through these categories. As Foucault (1977) notes, categories are self-regulating; they keep people in the places assigned to them via social norms and discourses. Students' testimonies have shown that the provision of education to disabled people based on categories and limited understandings of impairment can be inadequate and unfulfilling. The categories of special needs and disabled are not neutral technical terms but politically loaded ones which imply certain norms, values and assumptions and reveal particular ways of understanding impairment and disability (Benjamin, 2002; Race 2002).

In the context of HE, assumptions about disability and impairment are often made without reference to the experiential knowledge of disabled people. To create and perpetuate a category of disabled without examining or questioning able-bodied privilege hides the systemic and in-built exclusion of disabled people in education, historically and in contemporary contexts, as indicated in this section. As Benjamin (2002:124) notes:

> The governmental procedures that are invoked in relation to 'special needs students' are in many ways hyper-rational ones. The Code of Practice appears to construct a techno-rationalist reality in which struggles of any kind can be objectively qualified and quantified, and a solution found. If only we get the targets right, and the provision right, then the 'special needs student' will make the kind of linear progression that rational, developmental modes demand. And so difficulties in learning are measured, mapped and evaluated and remedial measures put in place to alleviate the effects of those difficulties.

Placing disabled students into what Benjamin (2002) calls hyper-rational or techno-rational categories, as it is carried out in schools in England, reduces experiences of impairment into something which is knowable, measurable and quantifiable. As evidenced by the students I interviewed, this experience of categories is foundational in that it sets the stage for their future encounters with able-bodied people in education and with the world at large. Achieving inclusion in education requires much more than is currently being done (hooks, 2003). In the next section, I explore how students navigate disabling attitudes, values and practices in HE, where those institutions' understandings of impairment are informed by notions of special needs.

(b) Navigating the Disabling Attitudes, Values and Practices of Higher Education

> Micro-social descriptions of our common sense practices are essential for those who want to take a macro-economic perspective…Ideological hegemony, as part of the actual workings of control, is not something one sees only on the level of macro-social behaviour and economic relations; nor is it something that resides merely at the top of our heads, so to speak. Instead, hegemony is constituted by our very day-to-day practices. It is our whole assemblage of common sense meanings and actions that make up the social world as we know it, a world in which the internal curricular, teaching and evaluative characteristics of educational institutions partake (Apple, 1995, in Benjamin, 2002:10).

This section of the chapter deals with micro-social processes, revealing what students say about their day-to-day experiences – from the classroom to every other aspect of student life. While my analysis of HEIs in Chapter 5 revealed the importance of the particular, of space, place and institutional culture, this section places student testimonies in the broader context of how SENDA is playing out in HE and considers questions about the importance of specificity in law and legal processes.

Building on the last section, I extend my analysis of the role of categorisation in underpinning the experiences of disabled students in HE. If categorisation has been infused into educational policy and practice, then it has also been mapped onto place and space, as evidenced by disabled students' lives (Dwyer and Bressey, 2008). The stories and encounters my interviewees shared suggest that HEIs are places which conceptualise disability and special needs in very particular ways.[10] HEIs, in terms of their physical environments, as well as their educational and social practices, have historically been created with little or no attention to impairment.[11] The interviewees reveal – through their daily encounters ranging

---

10   See Chapters 4 and 5 for more on this.

11   In fact, HEIs have traditionally been created with minimal inclusion of 'difference' of any kind – whether based on race, gender, sexuality, nationality or social class. See hooks (1994).

from trying to access inaccessible buildings to facing outright hostility – the underlying values, assumptions and attitudes of the institutions they attend.

A geographical understanding of the daily lives of disabled students is important because, as Imrie (2000:6) notes,

> For geographers, a paramount concern is with documenting the spatial patterns and processes underpinning such social relationships, traditions and lifestyles, and identifying some of the causal mechanisms involved in the production of diverse geographies.

In this section I bring to light examples of how categorisation is manifested in HE practices, and share students' stories about how they negotiate the spaces of their HEIs. A recurring theme for disabled students is the inescapability of the category. As SENDA is being interpreted and implemented in HEIs, some institutions are changing the scale and scope of categorisation within their practices. For example, students with dyslexia at Byron University must take several tests in order to be 'diagnosed' as having dyslexia so that they can then qualify for the accommodation of their impairment. The testing process is tedious and bureaucratic, and aimed at ensuring that only students who 'deserve' this diagnosis receive it. As Byron's disability officer noted:

> Some subscribe to the idea that if you encourage people...they will take advantage of special accommodations; it might be seen as somehow being unfair or somehow encouraging people who otherwise would not deserve them (Beverly, Disability Officer, Byron University).

One postgraduate student at this institution, Albert, said that the institution's way of conceptualising dyslexia, and providing uniform accommodations for students with dyslexia, is not necessarily helpful to him, and reveals a lack of engagement with, or understanding of, his particular needs. In her book *Teaching to Transgress: Education as the Practice of Freedom*, hooks (2002:29) emphasises that in her models of inclusive education, students have to be seen "in their particularity as individuals," and that progressive education means the active inclusion of difference and individuality in the academy. Indeed, it is this lack of attention to their individuality, a theme highlighted in the previous section of the chapter, which continually occurs in disabled students' experiences of higher education. Albert's experience of receiving uniform accommodations was also echoed by other students. For example, Chris said that his university has prioritised its responses to SENDA in very specific ways for groups of students with different impairments, and that their main concern is to cover all bases minimally, and to avoid litigation:

> Byron University has put dyslexia way down on the list. College is really worried about things they can get sued for, like ramps. As for dyslexia, they feel they've

> done enough...they don't think they can get sued for dyslexia but they've paid lip service (Chris, a student at Byron University).

As someone identified as having dyslexia, Chris felt that his educational needs were not being met. Yet, from the university's perspective, they were. The university classified Chris as having dyslexia and apportioned certain provisions which were thought to provide equal opportunities to all students in standardised ways. This notion of apportioning provisions is in keeping with medicalised definitions of disability which are central to the contemporary welfare state in Britain and other Western countries (Reinders, 2000; Stone, 1985). As Reinders (2000:72) says, "labelling occurs in order to see whether people fall under particular regulations or should have access to particular services." Chris's comment on lawsuits illustrates the increasing trend of HEIs attempting to avoid legal entanglements rather than meeting students' needs in engaged and meaningful ways. It is what Gewirtz (1997:11) calls the "triumph of the antidiscrimination principle" – the idea that all areas of social life can be fixed using legislation such as SENDA. As he says, "originally, nondiscrimination meant an obligation to treat people the same... today, nondiscrimination is coming to mean accommodating differences" (ibid.).

Like Gewirtz (1997), some disabled students think such faith in antidiscrimination law is misplaced. For example,

> Roger: I am sceptical of disability discrimination legislation...ultimately it's more important that someone's attitude changes so that they see there's no reason to have those attitudes.
>
> FV: Why are you sceptical of disability discrimination legislation?
>
> Roger: I see it as attitudes rather than anything else. I don't think you can change attitudes with legislation. It's about people's experiences, and experiencing disability firsthand to overcome prejudices (Roger, a student at Cromwell University).

Likewise, Chris felt that despite the existence of SENDA, disability discrimination is difficult to eliminate because it is deeply embedded in his institution's values. He said:

> It's the same as institutionalised racism because they're so unaware of it, how can they even begin to solve it? (Chris, a student at Byron University).

If disabled students feel that SENDA alone cannot end disability discrimination then how can equality for disabled people be achieved? Gewirtz (1997) has a solution: rather than the valorisation and reification of difference implicit in current antidiscrimination law, institutions must have meaningful engagement with that difference. Razack (1998) and Foucault (1977) would call for an unmasking of the power relations of those differences and would ask us to examine how notions of difference are produced, rather than simply slotting students into categories.

As demonstrated by Chris's story, uniform allocations and provisions for impairment according to category are not always practical or useful for disabled students. Peter, a student at Cromwell University, says that the academic curriculum was not sensitive to his particular needs as "the main problem is that it's all essay-based and I cannot write or spell." As Chris said: "it's such a variable thing; everybody learns and writes at different speeds." Each student is indeed different. Not all students read, write or learn in the same manner, let alone at the same speed. This is what his institution fails to grasp. As Chris said:

> I feel I'm being penalised because of their lack of understanding about education...they can't get their heads around the fact that they're an educational institution...Ninety five per cent of them don't give a monkey's ass. This idea that they're not going to get sued is crap. It's not just about ramps; they can do so much psychological damage to someone like me (Chris, a student at Byron University).

Chris was deeply unhappy about the form and content of educational provision available to him. Despite his efforts (he said "this year I made a big fuss, I spoke to the tutors many times"), the university refused to move beyond the idea that all students with dyslexia learn in the same way. It is this refusal to conceptualise learning above and beyond the category that so frustrated Chris.[12] In the context of the institution he attends, Chris found it difficult to escape the category of dyslexia. Byron University both reifies this category and simultaneously fails to engage with the needs of individual students. This recalls Foucault's (1979) discussion about how liberal institutions can be instruments of domination.

For the university, the 'problem' resides in Chris's impairment, whereas for Chris, who is trying his best to succeed, the barriers to his success are due to the institution's policies and procedures. As legal scholar Martha Minow (1991) notes, when difference is thought to reside in a person, and not in a social or institutional context, the institution can ignore its own role in producing that difference. As mentioned, Chris was frustrated that his university thought that his SENDA rights were fulfilled while in reality, his needs had not been met. Similarly, William, a student with multiple impairments, felt that the language of rights conceals existing inequalities. William felt that he was the target of discrimination for many reasons – including his impairments, his physical appearance, his speech, and his social class. He pointed out how the multiple forms of discrimination he experienced were not adequately addressed by a rights model. He said:

> You could be a female black disabled Muslim person and still have a right...I don't think that's actually going on in a world such as this, you expect a world

---

12   In fact he is so concerned that he has suggested that the Institute of Education conduct a study of the interactions between staff and students at his institution.

like this that everyone has their rights, but it's not like that...we have all these labels but...it doesn't work in a world such as this (William, a student at City of Bexley College).

Similarly, Razack (1998) writes about how ideas of rights can mask power relations and the limited choices people have. She says:

> The idea of rights, turning as it does on notions of individual freedom and autonomy, feeds the illusion that subordinate groups are not oppressed, merely different and less developed. Rights rhetoric, beginning with the idea that each person is free to pursue his or her own interests, masks how historically organised and tightly constrained individual choices are. The individual who has failed has simply chosen badly (Razack, 1998:24).

Chris is constrained – not by his own abilities – but by the limited choices on offer to him at the university, and the masking of these limitations by using a language of rights. For example, in his university's publication, *Disability Handbook: an individual's guide*, there are some internal contradictions in the document which suggest that despite the positive statement made about rights, there are problematic aspects to the university's approach to disability. According to the document,

> Byron University is committed to ensuring that prospective and current students, job applicants and members of staff have the right to be treated solely on the basis of their merits, abilities and potential without any unjustified discrimination on grounds of age, sex, disability... (2003:6).

The reference to unjustified discrimination suggests that there are forms of discrimination which are justifiable. Who has the authority to justify such discrimination? As the author of this document, and the ultimate holder of authority in its relationships with students, the university clearly holds this power.[13] Indeed, as another document from the same institution aimed at teaching staff asks,

> Must the institution provide the student all the academic adjustments requested? Not when the accommodation would fundamentally alter the nature of the program or when the academic requirements are essential to a program of study (i.e. compromise standards), or when the accommodation would cause undue hardship or burden (2003:36).

The university holds the power to create and enforce academic standards, and their guidance documents, both to teaching staff and to students, suggest that ultimate

---

13   SENDA acknowledges the role of institutional power. For example, SENDA allows institutions to treat disabled students less favourably or to not make reasonable adjustments in order to uphold 'academic standards', as defined by institutions (HMSO, 2001).

decisions about the accommodation of disabled students rest with the university. Another example of how issues of rights and power are manifested in institutional life is from the *Disability Handbook* mentioned earlier. In the opening vision statement, the university alludes to the rights of its students, staff and applicants. The statement also mentions the university's use of the social model of disability,

> Byron University recognises the 'social model' of disability: that disabled people are primarily disabled by the environment and/or others' attitudes in society, rather than by their medical condition (2003:4).

However, later in the booklet if refers to residence allocations for "students with mild disabilities" (p.20) and talks about "students with disabilities who are eligible for student loan" (p.23). These uses of language contradict the social model of disability as it is understood in the UK, and if the university were truly using the social model it would employ the term disabled students rather than students with disabilities. This insensitivity to differences in language suggests that while ideas about rights and the social model are employed, these ideas are not consistently applied in all the policies and practices of the university detailed in the handbook.

As demonstrated, despite the use of rights language, the constraints on individual choice are far more pervasive and deeply embedded than is immediately apparent (Foucault, 1980; Williams, 1997b). To understand where personal agency ends and narratives or broader discourses begin is not a simple task (Razack, 1998). Employing a Foucauldian (1980) analysis, Chris is not simply subjected to the university's power structures but is constituted by a set of power relations cast like a 'net' over how he thinks and views the world. It is important, as Chris's example illustrates, to remember that power is exerted not just overtly but covertly as well – in the context of the day to day and the banal (ibid.). To overcome this one must go beyond what Foucault calls the 'contract-oppression scheme' and instead see power as a "net organising how individuals are constituted in any one context" (Foucault, 1980:88–9)

Like Chris, Albert found it very difficult to break out of the category into which the university placed him. He found that he was compared to another student with dyslexia, and his tutor's expectations were based on such comparisons. He said:

> I was compared…when I got my feedback on an essay so he said 'Oh, very creative, like the other dyslexic student. Very creative; your thoughts are better organised than what he was.' Kind of like they're comparing, you know…it was a bit weird. It's like 'Oh, so if I'm dyslexic I'm supposed to follow that kind of pattern…' (Albert, a student at Byron University).

The idea that all students within given categories perform similarly is reinforced by policies and procedures, as evidenced in various documents procured from HEIs. For example, a SENDA strategy document from Byron University shows that various categories of impairment have been enumerated and each category is accompanied by academic accommodations and practices which are considered to be legally acceptable within the realm of SENDA. This shows that Chris's analysis is correct; his institution's strategy for complying with SENDA involves a reductionist approach of treating student impairment by category, and students therefore find it challenging to argue that their case may be different. Like Chris and Albert, Jacqueline was immediately slotted into a role when she started university: the angry young woman who constantly complains about wheelchair access. Literature in disability studies suggests that the 'angry disabled person' is an archetypal figure (Clare, 1999; Morris, 1991). This literature also indicates the difficulty for disabled people in being perceived by able-bodied people outside categories of disability (ibid.). Frederick, a student at Cromwell University, said, about the disability label that,

> It puts you in a big group, and there's all these other people out there struggling with it, and you're not (struggling) but you're in the same box with them – and it doesn't feel right (Frederick, a student at Cromwell University).

Jacqueline said that since her first day at university, she has been unable to 'break out' of the role assigned to her,

> I am perceived as being difficult…I went to see the Bursar, and he said 'don't come in, don't get comfortable, I know what you're here about' (Jacqueline, a student at Cromwell University).

While claiming to 'know' her so well that he did not want to even speak with her, the Bursar actually knew very little about Jacqueline's specific situation. She was living "in a second floor room with a lift that had been broken for eight years." Literature on the experiences of disabled people in the built environment indicates that disabled people have faced inaccessible environments for decades and even centuries (Doyle, 2000; Fincher and Jacobs, 1998; Gooding, 1994; Imrie and Hall, 2001; Stiker, 2000). Jacqueline said she was shocked when she came to college and found out where she would be living:

> I had the most unbelievable problems in terms of college and the university not being able or willing to accommodate. I told people about it. To start with they were unsympathetic until they realised I was on the second floor (Jacqueline, a student at Cromwell University).

She says that people had just assumed she was complaining because she was in a wheelchair, without realising that she was stuck on the second floor, literally trapped. She noted:

> For four weeks I had to call college porters; I had to be carried to go up and down the stairs…a lot of staff have a huge problem relating to how other people

feel...people see disability as about physical disability – the solution is that
porters can take you down. Disability discrimination occurs not just in terms of
access on the physical side but also on the mental side (Jacqueline, a student at
Cromwell University).

To Jacqueline, feelings and emotions are important. She wanted empathy, and to be
seen first and foremost as a human being, not an angry young woman in a wheelchair.
While the law can facilitate a physical solution – carrying someone down the stairs –
it does not in its current form facilitate the fulfilment of the emotional needs of
Jacqueline's situation, as she pointed out. So much of what she feels in her day-to-
day life is experienced above and beyond the legal realm of disability discrimination
as defined in SENDA. As Fitzpatrick and Hunt (1987:16) note,

> It is much easier to identify classificatory types which stand in direct connection
> with institutions...more difficulties are involved in the classification and
> characterisation of affectual and interpersonal relations.

Highlighting the importance of empathy in interpersonal relations, Jacqueline noted:

> It's almost as if they're not willing to put themselves in the place of someone
> with a disability. They are unable to imagine that they would ever have a
> disability themselves... According to the DDA, having a policy of carrying
> people in where lifts cannot be built is acceptable, but what about the person's
> dignity? (Jacqueline, a student at Cromwell University).

What Jacqueline's story illustrates is that beyond issues with the physical space
(the inaccessible second floor room) or the legal solution/accommodation made
(arranging for porters to carry her up and down the stairs), there is a domain of
human relations which is attitudinal, and which slips beyond the grasp of law and
legal process. As Jacqueline's story conveys, the realm of dignity, emotion and
empathy seems to escape the scope of the DDA, even as its minimal requirements
are being met. An example of this can be found in a strategic paper written by
a tutor at Byron University entitled *Disconnected for Connected Learning*.[14] In
the paper, the author differentiates between meeting the requirements of SENDA
versus what she calls the spirit of non-discrimination,

> Many UK universities have set up special IT labs, exclusively for use by
> disabled students and these facilities have demonstrated what can be achieved
> by deployment of appropriate assistive technology. For the most part such
> special IT labs have been set up as a facility separate from IT labs used by other
> students. However, in the true spirit of non-discrimination, the requirement is to

---

14   The paper suggests ways in which SENDA can be implemented with regards to
issues such as specialist assistive technology and communication networks.

embed the support technology in the regular student provisions, so that disabled students are accommodated in mainstream teaching and in one sense obliged to work in separate facilities (2002:2).

These separate labs indicate how an HEI itself differentiates between meeting SENDA's requirements – providing separate assistive technology labs for disabled students, versus being truly inclusive of disabled students in mainstream computing labs – a more laudable goal, but one which perhaps few institutions are achieving. Clearly, inclusion requires more than law and legal processes. As Roger's story indicates, much of what disabled students experience in negotiating HEIs as spaces and institutions has to do with attitudes, not physical structures or legal requirements. Roger attended the same university as Jacqueline but in a different residential college, and his experience was entirely different to hers. He noted:

> When I was admitted to the college, they were wanting to listen, to understand, there was no point trying to second-guess me. Whatever I requested, I got. They took out a cupboard and converted it into a toilet and shower. They installed a ramped remote control door. All of this was done prior to my arrival (Roger, a student at Cromwell University).

How is it that two students at the same university had such vastly different experiences?[15] Just as no two institutions are the same, there is variation within institutions.[16] As Gregory et al. (1994) note, space is influenced not only by physical factors, but also by the realm of the social and mental; space is both imagined and constructed.

Navigating the same university, one finds contrasting values, attitudes and practices depending on where in the university one is located and with whom one interacts. People do not interact with each other in uniform ways but rather bring their own personalities and experiences into human relations (hooks, 1994). This richness and complexity eludes categorisation, and likewise the experience of being disabled in HE eludes easy solutions by tick-boxes and standardised provisions.

This is an illustration of what Schuck (2002) calls a limit of the law – a complexity which the current legal system cannot grasp or include. Yet this is just what many HEIs are doing – indeed the guidance notes for SENDA, entitled *Estates: the Disability Discrimination Act Part 4,* suggests that access audits are a useful exercise for HEIs to help create accessible facilities on their campuses, "most institutions will find the first step in this process is to conduct an access audit of current provision" (DRC, 2002c:7). However, as literature indicates,

---

15   In fact, during Roger's time at university, SENDA was not yet in force, whereas during Jacqueline's time, it was.

16   For a detailed discussion of this, see Chapter 5.

approaching issues of accessibility from an 'audit culture' perspective may not be enough to resolve these issues (Imrie and Hall, 2001).

Roger's story illustrates that navigating their way through HEIs is challenging for disabled students and that geographies of difference occur in institutional settings. Despite the meaningful engagement that college staff had with Roger, he found himself constrained when he left college premises and went to general university sites such as the library. He said:

> The library was upstairs, so I got people to help me, we would go in groups to the library. I didn't have a problem with being carried upstairs, that was just part of my lot (Roger, a student at Cromwell University).

Roger's acceptance of his having to be carried upstairs as part of his "lot" has Foucauldian (1979) overtones of the self-regulating liberal subject. Likewise, Frederick felt grateful for the assistance he received at Cromwell and said, "a lot of the support I get seems like a bonus; I'd probably be struggling without it." For Roger, even with the college's many positive steps to be inclusive of him, he felt at times that he had to accept his lot. It is as if there was simply no way he could escape the category of 'wheelchair user' when he left the safe confines of his college to use the university library. This is not inherently due to Roger's impairment, but has to do with the way the university's space has been conceptualised and created – both in terms of the physical and attitudinal environments.

It is almost as if Roger's identity changed depending on his spatial location. In his college residence, where he was fully included, he felt "happy," was "very popular," and said his room was "a nice place to hang out." Outside this safe environment, Roger had to accept his lot as a disabled person and "depend on people's charity." This idea that identity changes in relation to space is supported by Armstrong's (1999) emphasis on the embodied and spatialised nature of identity (in Benjamin, 2002). Davies (1997) also employs a notion of identity which goes beyond traditional rationalist ideas, calling it the 'self-in-process;' a fluid, ever-changing contingent conception of the self (in Benjamin, 2002). As she explains,

> The subject of post-structuralism, unlike the humanist subject, then, is constantly in process; it only exists as process; it is revised and (re)presented through images, metaphors, storylines and other features of language, such as pronoun grammar; it is spoken and respoken (Davies, 1997, in Benjamin, 2002:16).

I find the 'self-in-process' a useful way of thinking about identity. As these students' testimonies indicate, not only is there great variation within the spaces of HEIs, but their comfort levels and sense of identity also change in relation to these spaces. I conclude this section of the chapter by trying to assess, overall, the impact of SENDA on these students' lives as they make their way through spaces of HE. While all the students interviewed are aware of SENDA, they do not think the legislation alone can end discrimination against disabled people. Roger, for example, who studied law and has a profound knowledge of legal matters, does not think SENDA has much hope of changing people's attitudes, which he sees as the primary source from which discriminatory behaviours and practices emanate. He does not see this as a particular limitation of SENDA itself but of law and legal processes in general.

Jacqueline, who was also sceptical as to whether SENDA could make a difference,[17] pointed to the multiple forms of discrimination she faced on a daily basis as evidence that law and legal processes alone cannot change her life. I include a large section of her narrative here to provide a flavour of a typical day for Jacqueline, uninterrupted:

> On my average day…it takes 40 minutes to go down the stairs and to go back up, which I do at least twice a day. Then, for me to get to lectures or tutorials, I have to take a taxi, which usually involves at least a 20 minute wait, even though I have booked the taxi in advance – and I usually take a taxi 4 times in a day.[18] College asks for detailed list of all my taxi expenses and where I've been – this delays things, my getting around, getting reimbursed and so on. All this time is taken up by disability-related crap so I can't do social things.[19]

When asked how she got on with fellow students, she responded:

> The first time I entered the department, they completely missed out on me, didn't talk to me, because of my height level – they were all standing and I was in my wheelchair. Sometime it's like I'm not even there…which often I'm not. For example, when there was a fire strike, the lady at the door stopped me from going into my lecture because they didn't have a procedure for evacuating wheelchair users from that building so I missed that lecture for 7 weeks. It's not just issues of physical access – I also get discriminated against because I have two disabilities. People found it really difficult to understand that there are two things…when I first started, College were really distrusting, they wanted to know all the details, including private medical information. They want all the information so that they can make a decision for me. They don't trust letters from my GP. I had a brain tumour in 1998 which causes me to experience tiredness. My Local Educational Authority (LEA) assessed me for disability student allowance and said I should be allowed extra time on tests and rest breaks, but college was

---

17   Though she does hope that students file lawsuits when and where possible.

18   Her College uses a company which is being sued and they can't find drivers, because drivers are not being paid (because of mounting legal costs). Her taxis were taking up to an hour to arrive. When she complained, they said 'Rachael, you have to book in advance,' they didn't believe her. On one occasion, knowing of the delays, she allowed 1 extra hour before supervision meeting, but still arrived 1 hour and 20 minutes late.

19   She is not a member of the disabled students' group. At the beginning of year they were present at the societies' fair. Unfortunately it was only one day and she missed it.

weird about it. They asked me, 'who's going to pay for an invigilator?' even though clearly college should. The LEA just said I should be granted rest breaks if needed, but did not specify a time. Neither did neuropsychologists. Despite the advice from my GP in a letter to college, I got an email from college with half the amount of time the GP suggested. For example they wanted my loo breaks to be taken out of exam time. They said to me 'why don't we time it – the first five minutes you're in the loo are fine, after that we deduct it from your exam time.' They were so nitpicky and so specific. They don't seem to accept that I'm the best judge of what I'm able to do. They are very controlling (Jacqueline, a student at Cromwell University).

Finally, Jacqueline offered her view of SENDA's efficacy:

Change is hard to achieve in this university. The University is very resistant to a change unless it gets backed into a corner. Lecturers who teach here are so arrogant...Cromwell University does have a tendency to ignore what the rest of the world does. They ignore the fact that public services have to come up to standard. British Rail caters to my disability better than Cromwell University. When I've said "Do you realise this is illegal?" they've said, "Well, this is Cromwell University" (Jacqueline, a student at Cromwell University).

As Jacqueline's story indicates, disabled students' experiences of HEI spaces are punctuated by subtle forms of exclusion as well as blatant expressions of ableism, prejudice and hostility which can impact all aspects of their lives. As noted earlier, Chris is also sceptical of SENDA's ability to improve his life at university. Specifically addressing the issue of power, and how tutors can exercise it in different ways, Chris says that one of the ways tutors exert their power is by not responding to student requests. For example,

The department administrator, who is very nice, told me to write to my personal tutor about my issues. I did and he didn't respond. Useless (Chris, a student at Byron University).

Chris postulated that the tutor in question did not respond because he knew that there would be no repercussions for not doing so, and that Chris would have no other recourse. Indeed, guidance material on SENDA points out that it is not a compliance-based law and therefore the only legal recourse for students is to challenge HEIs in court or through the DRC mediation process (Byron University, 2003). This section of the chapter has indicated that the category of disability has been mapped onto the spaces of HEIs and that disabled students want to experience higher education above and beyond this category. Finally, this section has also shown the limits of SENDA, and indeed of law and legal processes, in reaching all aspects of human life and experience. The concluding section will summarise the main themes of the chapter and provide suggestions on starting points for inclusion in higher education.

(6.3) Conclusions

This chapter has sought to enhance understandings of the richness and complexity of disabled students' lives, and to place their stories in broader contexts of law, the categorisation of disability, and spaces and places of higher education. The chapter employed Razack's (1998:36) notion of storytelling to inject the daily-lived experiences of disabled students into "established" and "dominant knowledge paradigms." Working through themes about the experience of categorisation, and the ways in which categorisation underpins how disabled students navigate the spaces and places of HEIs, it illustrated the diverse ways in which the categorisation of disability has material impacts and consequences in their day to day lives.

These experiences – ranging from inclusion to exclusion, from subtle harassment to overt hostility, and from the physical inaccessibility of HEIs to the exclusivity of curricula and conceptions of education – uncover how challenging it is to capture peoples' lives into the limited categories of discrimination envisaged in the DDA and SENDA. All of the students interviewed spoke of the difficulty of changing attitudes using legal frameworks, and, to them, it is in the attitudinal realm where the most significant changes are needed.

Using storytelling as a starting point for inclusion, this chapter illustrated the different ways in which disabled people experience the world and navigate spaces and institutions (Clare, 1999; Razack, 1998). It is hoped that these findings will have relevance for, and feed into debates about, how to redress disability discrimination. Specifically this critique engages with the main antidiscrimination clauses of SENDA, which state,

> (I) HEIs must not treat a disabled person less favourably than a non disabled person for reasons related to his or her disability, without justification, and,

> (II) HEIs will be required to make reasonable adjustments to ensure that a disabled student is not placed at a substantial disadvantage (HMSO, 2001).

From the interviewees' daily experiences it is clear that even with SENDA in place, disabled students are treated less favourably, such measures are being justified, the definition of reasonable has been left up to HEIs, and disabled students are often placed at a substantial disadvantage. Redressing disability discrimination in HE will involve both working within existing legal frameworks to include more experiential knowledge of disabled people, and working above and beyond the law to achieve inclusion that is not limited to merely treating a disabled person less favourably, but to actually including and embracing them as an integral part of higher education. Relying on alone SENDA ignores the complex power dynamics within HEIs, and within the legal system, which have deeply embedded ableist attitudes, values and beliefs, and thus put disabled people at substantial disadvantages systematically (Foucault, 1979). Ultimately, HEIs need to move beyond the paradigm of non-discrimination towards full equality and active inclusion.

PART III
Conclusions

Chapter 7

Law's Role in Creating Inclusive Environments for Disabled People

(7.0) Introduction

As Blomley (2000) notes, law is conceptualised by many legal authorities as abstract and removed from the particular circumstances of material life. He writes that "legal relations and obligations…are frequently thought of by courts and other legal agencies as existing in a purely conceptual space, with little recognition of the spatiality and diversity of material circumstance (Blomley, 2000:53). This book investigated the experiences of disabled people in relation to the DDA with the aim of exploring the limits of this conceptualisation.

As argued throughout the book, law is not purely abstract or removed from everyday life. Indeed, the purportedly autonomous neutral liberal legal subject is predicated upon particular experiences of, and assumptions about, material life (Pue, 1990). This conceptualisation of the legal subject has tended to favour able-bodied norms and to disadvantage disabled people (Gooding, 1994). Furthermore, as indicated in the chapters on higher education, and by various theorists, law is implicated in the categorisation and control of disabled people (Benjamin, 2002; Foucault, 1979; Race, 2002; Razack, 1998).

This chapter considers two of the main themes of the book: liberal legalism and categorisation. In concluding, this chapter explores alternative paths to inclusion and reflects on areas of further inquiry. The chapter is divided as follows: (7.1) Law and its liberal limitations, (7.2) The tyranny of the category, (7.3) Alternatives to law, and (7.4) Areas of further of study.

(7.1) Law and its Liberal Limitations

Attempts to evaluate and assess the impact of a policy or law can be riddled with complexity and ambiguity. As Mansell and Ericsson (1996:ix) argue: "the true consequences of major changes in any sphere of public policy are revealed in a frustratingly slow fashion…given the vagaries of execution and the inevitable slippage between aspirations and application." Investigating key actors' and disabled peoples' responses to the DDA, illustrates how vast and complex the territory of disability discrimination law is, and therefore, how modest and any claims about their value must be. This book has aimed to shed light on how institutional actors and disabled people are influencing, and being influenced by, the law.

One of its main themes of is the importance of specificity and context. Law and legal processes are often created in abstraction from, and are therefore not sensitised to, people's particular experiences and contexts. The research findings of this book indicate that the DDA, in its liberal legal modes, is idealistic and removed from everyday realities of disabled people and their particular experiences and geographies. Critical legal geographical ideas of context, specificity and space are important if the DDA is to achieve its goals (Clark, 2001). As noted in Chapter 2, the DDA (UK), DDA (Australia) and the ADA must be viewed within current socio-political and legal contexts in order to be fully understood. Moreover, the according of 'rights' within legal systems which are exclusionary or hostile to disabled people, can infringe upon these rights, or, in some cases, even harm them (Roulstone, 2003).

As noted in (7.0), the liberal legal subject conceptualised in the UK, and Australia and the USA, is generally presented as neutral, individualistic and self-serving. However, this is also a very particular type of subject, and as evidenced in the built environment, it is assumed that the 'norm' for this subject is able-bodied. As discussed in Chapter 3, the DDA is considered problematic by disabled people because of its normalising stances, for example in its medicalised definitions of disability. Evidence suggests that Part 3 maintains able-bodied norms and societal structures and simply aims to make these more accessible to disabled people, and only in ways which are 'reasonable.' The DDA's vision of 'access' is limited one, and must be expanded and broadened if inclusion is a goal of the legislation. As Harvey (2000:185) notes,

> Materialized Utopias of process cannot escape the question of closure or the
> encrusted accumulations of traditions, institutional inertias, and the like, which
> they themselves produce.

The DDA does not allow for an 'opening up' of corporate hierarchies, decision making authorities, or 'closed' power structures which favour able-bodied norms and have traditionally excluded disabled people (Birch, 1998; Dearlove and Saunders, 2000). Laws and rights may be good starting points for inclusion but they need to be implemented in enabling environments. For example, material outcomes for disabled students depend on outdated able-bodied assumptions, perceptions, attitudes and values. This is not surprising given the way older higher education institutions such as Cromwell operate in outmoded ways and in some cases are even proud of this 'traditionalism'.

The DDA is therefore limited because of its medicalised definitions of disability, and the fact that its enforcement relies on disabled people to fight cases after discrimination has occurred. As noted in Chapter 2, as in America and Australia, disabled people taking up their cases in the UK find many obstacles. These include a judicial system which is biased in favour of able-bodied people due to lack of understanding of disabled peoples' realities, lack of funding for legal representation, backlogs of cases, and processes of going through antidiscrimination claims which can be dehumanising, demoralising and damaging (Roulstone, 2003).

In the context of higher education, I demonstrated in Chapters 4, 5 and 6 that SENDA is being interpreted and implemented in keeping with the managerialisation of the sector (Newman, 2001). It has been fascinating examining this sector at this particular time as it is undergoing significant shifts and facing major challenges in the form of top-up fees, increased diversity i.e. more international students and scholars in the UK, and greater competition from HEIs abroad. How SENDA fares within the context of the larger priorities of senior HE managers in such a pressurised and competitive environment is interesting to observe, and a more detailed and long term study in this area would be interesting, as further discussed in (7.4).

Interviews with disability officers and other members of staff indicate that even with SENDA's existence, those who are not committed to equality and inclusion can find ways of subverting legal processes. For example, one disability officer noted:

> I think in this field it's very easy to pay lip service to doing the right thing, and to look good. But you might not actually have the substance behind it and I'm very loathe to encourage that kind of service/tick box approach, which means you can say we've done that we don't need to worry about that, we've got a policy, we've got this, we've got that. Without actually having the *structures*, and the *support*, and the *awareness* to do that (emphasis added).

As indicated, there are other criteria, beyond the law, which are necessary in order for inclusion to materially benefit disabled people (Brooks, 1997; Razack, 1998; Young, 1990). In addition to the DDA, inclusion can benefit from: (1) the availability of sufficient funding and other resources for increasing accessibility (in the broadest sense, not just physical access) for those who need it, (2) the availability of knowledge and information about how to increase access on a large scale, (3) broad awareness-building and educational campaigns on the legislation, its philosophies, and how it intends to empower disabled people, and (4) institutional structures (in governments, quangos, HEIs and corporations) which are egalitarian and include disabled people at all levels of decision-making. From my research, I have found no indication that any of these four processes are occurring, although I know that people in the DRC are working hard to raise the legislation's profile and to target these four areas despite the organisation's limited resources, and its possible merger with the EOC and CRE.

In this section of the chapter I have argued that law and legal processes have limited ways of looking at peoples' lives and their contexts. Law is not necessarily effective at relating to attitudes, values and other subtle forms of discrimination which disabled people encounter on a regular basis. My research conveys the experiential knowledge and feelings of 'what it is like' to be a disabled person in higher education, trying to get in, get on, and navigate their ways through the sector. In excavating the materiality of being a disabled student, these experiences are linked to broader political programmes – of managerialisation, medicalisation, categorisation and liberal legal goals. I now discuss disabled peoples' experiences of categorisation in the next section, (7.2).

(7.2) The Tyranny of the Category

I have argued that the categorisation of disabled people is presented as a neutral tool for 'managing' people and systems, but is in fact underpinned by particular assumptions and values. As noted in Chapter 5, HEIs like Cromwell can be very traditional, and institutional actors sometimes engage with issues of disability in ways which are in keeping with these values. For example, some disabled students at Cromwell told me that while all their requirements were met, they were done so in patronising ways, or that members of staff employed notions of 'care' when responding to their requests. At Bexley, which is closer to Barnett's (1990) typology of a radical institution, members of staff indicated that despite fostering notions of inclusion, negative attitudes towards disabled students still exist. Within both institutions there are examples of both inclusion and exclusion, and as both HEIs comprise multiple sites, there is much intra-institutional variation regarding responses to disabled students. However, regardless of which institution they attended, disabled students interviewed were emphatic that they disliked being categorised as disabled or having their identities associated exclusively with impairment.

In particular, these students welcomed academic and personal support services, but criticised the language of special needs, and associated it with ideas of segregation, isolation and patronising notions of care (Benjamin, 2002; Race, 2002). Furthermore, these students indicated that these limited understandings of disability and impairment can lead to inappropriate forms of support or even discriminatory attitudes and behaviours aimed at them. While SENDA's goals may be the inclusion of disabled students, its effects are that some universities are increasing their segregation, classification and categorisation of disabled students. For example, as noted in Chapter 6, one institution is increasing the bureaucratisation of dyslexia diagnoses in response to SENDA. Therefore, it is possible that the discourses of disability currently used in UK higher education, even post-SENDA, can actually lead to further segregation of disabled students, as noted in Chapter 5 with the example of separate computing laboratories for disabled students. Such actions can have the effect of calling increased attention to disabled people's 'otherness' and difference (Foucault, 1975; Said, 1978).

Some disabled people want to demystify the 'category' of disability, to take it apart, and to be engaged in a politics of re-categorisation, or of going beyond categories. For example, one interviewee told me that she refers to able-bodied people as temporarily able-bodied (TAB), as in her view, when we age we all acquire some form of impairment. She felt that the normalisation of able-bodiedness in society is so automatic, unseen, and taken for granted, that disabled people need to engage in radical political activism and to change what she sees as the 'monopoly' which able-bodied people have over language and other cultural norms. How can disabled people engage in such political and linguistic projects?

One way is through challenging discourses (Foucault, 1977). Throughout the book I have engaged with ideas of storytelling, examining discourses, and inserting disabled peoples' voices into broader debates (Clare 1999; Razack, 1998). The testimony from disabled people, particularly in Chapters 3, 5 and 6, points to the strong connections between liberal legal medicalised definitions of disability and the categorisation, stigmatisation and oppression of disabled people. In higher education, significant elements of disabled students' lives – such as where they live and how they study, for example – are influenced by their institutional actors' ways of understanding disability, and providing resources, policies and procedures based on these understandings. This book has demonstrated the connections between the ways in which disability is conceptualised and the material disadvantages faced by disabled people. If able-bodied people are to be more inclusive of disabled people, this will require not just legislation, but also attitudinal and societal shifts, and other ways of engaging in these important debates. Having considered the limits of liberal legalism and the tyranny of the category, in (7.1) and (7.2), I now move on to consider non-legal or supra-legal routes to inclusion.

(7.3) Alternatives to Law

> A critique is not a matter of saying that things are not right as they are. It is matter of pointing out on what kinds of assumptions, what kinds of familiar, unchallenged, unconsidered modes of thought, the practices that we accept rest... Criticism is a matter of flushing out that thought and trying to change it: to show that things are not as self-evident as one believed, to see that what is accepted as self-evident will not longer be accepted as such. Practicing criticism is a matter of making facile gestures difficult" (Foucault, 1988 in Olssen, 1999:113).

Chapter 2 demonstrated how the DDA was premised on utopic notions of all liberal subjects being free and equal, of ignoring the material realities and differences of disabled people. I have also argued throughout the chapter that disabled peoples' voices need to be heard as a starting point for inclusion. Beyond storytelling for social change, some theorists point out that the first step for creating a new and more inclusive world is to imagine and design one (Harvey, 2000; hooks, 2003). This is not a trite notion, but one with strong philosophical and theoretical underpinnings, and has been applied in practice, for example by the Brazilian Workers Party, as documented by Unger (in Harvey, 2000). For example, Lefebvre (1991) and Unger focus on social processes and institutional change through personal transformations (in Harvey, 2000). Unger avoids simplistic notions of utopianism by emphasising that alternatives should emerge out of critical and practical engagements with the institutions, personal behaviors and practices that now exist ... only by changing our institutional world can we change ourselves at the same time, as it is only through the desire to change ourselves that institutional change can occur (in Harvey, 2000:186).

Unger also proposes three main forms of empowerment which can help to overcome institutional power structures and be liberatory and transformative: (1) the opening up of social life to practical experimentation, (2) the strengthening of our "self-conscious mastery over the institutional and imaginative frameworks of our social experience", and (3) the cleansing of collective life of its capacity to entangle people in relations of dependence and domination (in Harvey, 2000:187). Unger's notion may sound simple but his commitment to transformation is unwavering and his philosophies have been successful in some contexts (ibid.). As Harvey (2000:186) notes, by changing our institutional practices we can change ourselves at the same time, "as it is only through the desire to change ourselves that institutional change can occur." Clare (2001) also believes this based on her own experiences as a disability activist and notes that the work of reconfiguring the world is often seen as changing the material, the external, but at the same time,

> Our bodies – or, more accurately, what we believe about our bodies – need to change so that they don't become storage sites, traps, for the very oppression we want to eradicate. For me, this work is about shattering the belief that my body is wrong (Clare, 2001:363).

Clare's view is compatible with disabled, feminist and anti-racist writers who connect the personal to the political (Oliver, 1990; hooks, 1994; Razack, 1991). In thinking of ways of working above and beyond the DDA, within and outside of legal realms, there are many possibilities. For example, some theorists feel that governments, either through law or through other tools at their disposal, should do much more to improve the lives of disabled people through one of the following ways which go beyond the current limitations of civil rights legislation: (1) creating laws which cover areas not addressed in the DDA such as social norms of disability and cultural portrayals of disabled people, (2) creating laws that fundamentally change the structure of society, moving beyond liberal ideas of social interaction and the distributive paradigm of justice (a paradigm which does not address subtle forms of oppression), (3) create laws which are not ideologically based on the individual-medical model of disability, and finally (4) the use of non-legal remedies beyond anything previously imagined in Western societies. Gooding (1994), in particular, makes the argument that the formal equality model has failed.

(7.4) Areas of Further Study

Given the limitations of time, resources and scope, this book is only a small part of a much broader field of inquiry into disability discrimination. Disability research would benefit with further analysis, in the following ways: (1) quantitative or qualitative studies on disabled students in education in the UK, starting at the GCSE level or younger, so as to find out about their pathways into higher education, (2) research into the career choices and employment experiences of disabled students upon leaving UK higher education, (3) research into whether the legal processes of the DDA, such as employment tribunals "disable significantly or long-term the integrity, and confidence of disabled complainants" (Roulstone, 2003), and finally (4) more research generally into interlocking systems of oppression within the UK, and using other western countries as points of comparison (Morris, 1996; Razack, 1998).

Understanding how interlocking systems of oppression work is important, because as mentioned in Chapter 2, disabled people face systematic disadvantages in society, and these are theoretically and practically tied to the oppression of all marginalised people. This is because disability cuts across all segments of society, regardless of age, gender or ethnic background (Clare, 2001; Morris, 1996). Ultimately, all of the above must be linked in terms of discourses of fighting oppression – otherwise you create space for prioritising one type of oppression over another, e.g. racist women, sexist gays and lesbians, homophobic disabled people, ableist people of colour, etc. (hooks, 1994; Razack, 1998; Young, 1990). Interlocking systems of oppression are important, but not in the way government ministers envisage them. By collapsing the EOC, CRE and DRC into the Equality and Human Rights Commission (EHRC), primarily for budgetary savings, rather than a true commitment to tackling interlocking systems of oppression, the government has compromised the ability of disabled people to access the legal system; the DRC had symbolic and practical resonance with disabled people which the EHRC has not inspired.

For my final thoughts I return to Razack's (1998) *Looking White People in the Eye*, a work which has loomed largely in my life, personally and professionally, since I first read it in 2002. In thinking about interlocking systems of oppression, Razack (1998) suggests that we challenge conventional ways of seeing women of colour and disabled women as suffering from 'double' or 'triple' discrimination, but rather we try to gain an understanding of these women's oppressed situations by examining the power relations that have led to discrimination. Razack (1998) also deftly turns 'the gaze' around and asks academics to look at themselves in the mirror. She challenges us to see that we are all implicated in the social, economic and political structures which oppress marginalised people.

Razack (1998) believes that solutions will entail more than creating laws or introducing programmes. Like Clare (1999), Razack (1998) notes that revealing how systems of domination and oppression work can help make it easier to dismantle them. This book has tried to demonstrate how systems of oppression work in HEIs in England and Wales, and how those disabled students who enter HE have used strategies to counter and resist such oppression. As shown, these struggles can occur on a daily basis for these students. As I end this book, I too feel that the road to inclusion is long and that there are many more struggles and much work to be done.

Epilogue: The Equality Act [2010]

Effective 1 October 2010, UK law pertaining to disability discrimination comes under the jurisdiction of the new Equality Act [2010] which effectively replaced the Disability Discrimination Act (DDA) 1995. The definition of disability employed in the DDA remains essentially unchanged in the Equality Act. Disability is now described as a "physical or mental impairment that has a substantial and long-term adverse effect on an individual's ability to carry out normal day-to-day activities" (HMSO, 2010).[1] The government has replaced the gender-specific reference of 'his' with 'an individual's.' Beyond this minor adjustment, which will not substantively affect legal outcomes, the new law has a range of ramifications for disabled people, which I examine as follows: (1) changes to the definition of disability and discrimination in the Act and, (2) the broader UK socio-political climate in which the Act is ensconced.

(1) Definitions of Disabled and Discrimination

The definition of a disabled person employed in the Equalities Act 2010 does not include a list of capacities that a disabled person's impairment has to demonstrably impact unlike the DDA 1995. This could theoretically make it easier for an individual to demonstrate that s/he is disabled and potentially enable tribunals to determine whether or not an individual's impairment has a legally significant effect. Conceptually, like the DDA, the Equality Act's definition of a disabled person appears to be based on the medical rather than the social model of disability. Beyond its similar definition, there are practical considerations regarding the potential efficacy of the Act. Most significantly, there is no indication of additional resources allocated so that individuals will be able to access the legal system; this could have significant impacts on the ability of disabled people to redress discrimination beyond definitional debates.

Indirect discrimination is now included in the Act, bringing the UK in line with legislation in the USA and Australia. In the context of the Equality Act [2010], this means that if someone is discriminated against because of their association with someone who is deemed to be disabled to be disabled under the act, such

---

1 Previously, disability was defined in the Disability Discrimination Act as follows: "A person has a disability for the purposes of this Act if he [sic] has a physical or mental impairment which has a substantial and long-term adverse effect on his [sic] ability to carry out normal day-to-day activities" (HMSO, 1995:1).

discrimination would be unlawful; this is also known as associative discrimination. Associative discrimination is perhaps a more accurate descriptor than indirect discrimination, which could be construed as discriminating against a group of individuals rather than against a particular individual. In practice, the inclusion of indirect discrimination in the ADA in the USA has enabled a series of class action lawsuits which may still not be possible in the UK.

One of the features of the Equality Act [2010] which was not present in the Disability Discrimination Act [1995], or its amendments, is an expanded list of categories of impairment which are no longer considered recognisable impairments for the purposes of the Act. One of these, for example, is a tendency to set fires (HMSO, 2010). One might ask how this particular item came to appear on a list of impairments which are not considered 'disabilities' for the purposes of the Act. This particular example, of arsonists, may have arisen from a consultation prior to the introduction of the Act, or there may be significant case law indicating that arsonists navigating the legal system routinely attempt to claim disabled status.[2]

One possible analysis of this particular condition finding its way onto the list of excluded categories has to do with increasing attempts by governments to entangle the social welfare state with the criminal legal system (Ward, 2011). This is not an accidental occurrence, but rather part of a wider attempt of the present and previous governments to associate the improper claiming of benefits – or the perceived over-reliance of individuals on the state – as morally reprehensible and deserving of legal penalties and social stigma. An example of this can be seen in David Cameron's statements following the August 2011 riots, in which he said these occurred due to the "twisting and misrepresenting of human rights in a way that has undermined personal responsibility" (ibid.). Immediately following the riots, housing minister Grant Shapps introduced a proposal to the backbench business committee that individuals who are convicted of anti-social behaviour would be automatically evicted from social housing (Prochaska, 2011).

In particular, with regard to disability, there appears to be an underlying assumption that there is a need – either conceptually or for pragmatic reasons – to limit the number of people claiming to be disabled under the Act. This is in harmony with the wider legal climate in the USA over the past decade in which decisions arising from case law appear to narrow the definition of disability as much as legally possible within the confines of the Americans with Disabilities Act (ADA), as evidenced, for example in the U.S. Supreme Court case Toyota v. Williams.[3]

---

2 There is significant debate about the motivations behind arson given what Read and Read (2008) call the over-representation of learning disabled people among arsonists. Freud's theory of pyromania has been largely discredited since the 1960s but there are studies in the fields of mental health and psychology indicating relationships between the occurrences of arson and learning disability (Hall et al., 2005).

3 For more on the Toyota v. Williams case, see Chapter 2.

(2) Broader Socio-political Climate in the UK

In the national election held on Thursday 6 May 2010, Britons' votes resulted in a hung parliament. The result was a coalition government formed of two different political parties: the Conservatives with 306 seats, led by David Cameron, who would serve as Prime Minister and the Liberal Democrats with 57 seats, led by Nick Clegg, who would serve as Deputy Prime Minister. The coalition government, as it has come to be known, was meant to represent a philosophical balance between the Conservative prioritisation of fiscal austerity and the Liberal Democrat focus on social justice, albeit it in a liberally-conceived, individually-focused society, in keeping with the utilitarian principles of classical liberalism (Mill, 1869). It remains to be seen how or whether this balance will be achieved; early indications are that, like the previous government, this one is engaged in internal debates about appropriate governing philosophies and ensuing courses of legislative action.

As is the case with the Equalities Act [2010], a change of government ensued around the time that the Disability Discrimination Act [1995] came into force; the UK government changed from Conservative to Labour in 1997. Paradoxically, the Labour government inherited a law which some felt was philosophically aligned to the interests of the apparently natural constituents of Conservatives: businesses, the wealthy, and multinational corporations.[4] The Labour government then amended the law to bolster its efficacy, e.g. the Special Educational Needs and Disabilities Act (SENDA) 2001 and the Equalities Act [2006]. Labour's record on equality is mixed, however, especially if one closely examines the output of Equalities and Human Rights Commission (EHRC) both conceptually and in terms of the practicalities of resource allocation.

To take just one example, in June 2011 Trevor Phillips, the chair of the EHRC, stated that "our business is defending the believer" (Copson, 2011). This was in response to the cases of Lillian Ladele, a registrar who refused to fulfil her duties because of her opposition to same-sex partnerships, and Gary McFarlane, who refused to treat gay couples in his role as a Relate counselor. As one writer points out, Phillips' statement is legally inaccurate: "the role of the EHRC is actually to protect the rights of both religious and nonreligious people" (ibid.). The commission announced in July 2011 that it was applying to intervene on behalf of these two Christians, along with two others who alleged discrimination in the workplace, and was taking these cases to the European court of human rights. At issue is the alleged prioritisation of the rights of religious individuals to maintain their beliefs in the workplace over the rights of gay individuals and couples to be protected from discrimination.

The EHRC announced its support of the cases of the four Christians and justified this support because it views equality law as it is currently being interpreted, as insufficiently protecting freedom of religion or belief. However, shortly thereafter,

---

4   This view was espoused by multiple civil servants interviewed.

the commission issued a clarification that "under no circumstances would [it] condone or permit the refusal of public services to lesbian or gay people" (quoted in Copson, 2011). The compromise position appears to be that these Christian individuals could be entitled to refuse service to gay people as long as alternative counsellors or registrars could be found who would be willing to serve as replacements for the religious individuals. Copson (2011) argues that "this would be to argue for a retrograde step in English law" and points out that in the case of Lillian Ladele, a precedent was set which recognised the incontrovertible rights of lesbian and gay people to be protected from refusal of service, and that this refusal could be construed as discrimination.

Of wider concern is the possibility that the rights of certain categories of individuals could take precedence of over the rights of others, thus failing Kant's categorical imperative,[5] and a trajectory foreshadowed by an interviewee at the now defunct Disability Rights Commission, who pointed out that when the various equality bodies in Northern Ireland were merged, the resources and attention dedicated to the rights of disabled people were far less substantial than they had previously been. As Copson (2011) argues, "the priority that it reveals the EHRC is giving to these cases...is almost more shocking" than their substantive content – i.e. that the rights of religious people should legally trump the rights of gay people. This indicates a wider point about the latitude the EHRC has been given, the prioritisation of the rights of certain constituencies or categories of individuals over others, and even about its chair, hardly a non-controversial figure himself.[6]

Given all of these controversies, much debate ensues about the merits of legislative and judicial approaches to equality law[7] and to the existence of these

---

5   Kant argues that the moral worthiness of acts should be evaluated on the basis that they be universally applicable: "act only according to that maxim whereby you can, at the same time, will that it should become a universal law" (Kant, I, 1785) trans. by Ellington J. (1993:30).

6   For example, in 2009, it was reported that "a cross-section of his commissioners were calling for his head" and that several of them chose not to renew their contracts with the EHRC, citing Phillips' outspokeness and insentivity on issues such as race and multiculturalism. See Muir, H. (2009), Trevor Phillips: a career in crisis, *The Guardian*, 28 July 2009.

7   For example, see Home Secretary Theresa May's lambasting of the Human Rights Act at the Conservative Party Conference in September 2011; she argued that the Act was interpret to allow an asylum seeker the right to stay in the UK because of his ownership of a cat. This was both inaccurate (Justice Secretary Ken Clarke publicly refuted May's claim) and inappropriate: there are dangers to the integrity of the legal system if the executive branch of government pronounces judgment on judicial decisions and undermines the credibility of judges' decisions and uses the realm of public relations and media stories to do so. Ward (2011) warns of the dangers of undermining the integrity of the legal system if the executive branch questions judicial decisions.

legislative tools themselves.[8] In the current climate, the coalition government finds itself in a similar position to the Labour government in 1997; it has inherited a law which appears to bolster equality and with which it ought to want to be associated. However, it may be philosophically opposed to the way in which equality is envisaged in this particular legislation. Furthermore, attributing a viewpoint or perspective to a government is an exercise which should always be undertaken with great caution. Current times call for even more circumspection; with two parties sharing power on one side of the benches of the House of Commons, one cannot definitively ascribe a single view to the government.

At times, however, hints emerge. For example, in late 2010, Conservative Home Secretary Theresa May stated that the government would consider scrapping the Human Rights Act and replacing it with a British-centred Act, rather than the current Act which is gravitationally pulled towards the European Court of Human Rights. At other times, May has mooted the idea of repealing key provision of the Equality Act, such as its provisions on socioeconomic equality (Monaghan, 2011). Others in the government - such as Deputy Leader of the Liberal Democrats Simon Hughes and the Liberal Democrat Equalities Minister Lynne Featherstone – have bemoaned stagnant or growing inequality in the UK and appear committed to combating it. The outcome for people who do not experience full equality, however measured, hangs in the balance.[9]

Indeed, given the August 2011 riots which spread across the UK in the wake of the shooting of Londoner Mark Duggan by police in Tottenham, concerns about equality, or a lack thereof, are very much at the forefront of socio-political discussions in the country today.[10]

When considering the material impacts of the current government's approaches to equality, in addition to the political rhetoric surrounding equality and human rights, financial considerations appear at the forefront. The overarching narrative of this government's legislative agenda, fairly or unfairly, is about "cuts." The UK government is committed to the cutting of billions of pounds from the national budget in order to eliminate the structural deficit by 2014/15 (Umunna, 2011). Local authorities state that these cuts will result in the closure or discontinuation of libraries, care centres, community enterprises, grants for social enterprises, education, health care and a whole host of other state-funded social apparatuses

---

8   May has also mooted the idea of scrapping the Human Rights Act when she stated: "I'd personally like to see the Human Rights Act go because I think we have had some problems with it." Source: Hennessy, P, "Home Secretary: scrap the Human Rights Act", *The Telegraph*, 1 October 2011.

9   For example, see the story on the eviction of travellers from Dale Farm: "Judgment Day", *The Economist*, 13 October 2011.

10   For example, Green Party members Caroline Lucas and Jenny Jones wrote a story entitled "To recover from the riots we must rebalance the inequalities of society," The Guardian, 22 August 2011.

which are to be taken up by the private and/or voluntary sectors, under the aegis of the Big Society,[11] or, simply to disappear altogether.

In line with this fiscal austerity, there is a concerted effort to remove as many individuals from the welfare rolls, through the limiting the Jobseekers' allowance, renamed the Employment Seekers' allowance, to a period of one year. A move is also afoot to apply more stringent criteria (a move also presaged by the previous government, in its elimination of the Incapacity Benefit) to the Disability Living allowance by increasing the waiting period for eligibility, limiting the length of its application, and reducing amounts paid out.[12]

Taken together with the new Equality Act, it is understandable why reactions from disabled people range from sceptical to outraged. For example, on 22 October 2011, large public gatherings were organised across the UK to protest the proposed changes to the Disability Living Allowance (Rose, 2011). These debates indicate that regardless of what equality rights are enshrined in law for the protection of disabled people, without the structural means by which to be enforced, or without adequate resources provided to those seeking to redress inequalities, laws are not fully effective.

Disabled peoples' concerns about these cuts feed into wider public sentiment in a time of economic insecurity and an almost Hobbesian focus on securing individual economic security and seeing others' entitlements, such as the Disability Living Allowance, as either unworthy or as resources which do not necessarily have to be provided by the state. In relation to the role of individual citizens vis-à-vis the state, this is the big question of our time. As many have posited, neoliberalism faces a deep existential crisis. Labour leader Ed Milliband, in his speech to his party conference in September 2011, argued for the creation of new models for regulating the economy which focus on social responsibility rather than on the unimpeded creation of wealth for individuals (Milne et al., 2011).

Alternative social visions[13] to neoliberal market capitalism have been few and far between and difficult to implement. Some ask what practical forms of social cohesion these alternatives can provide in times of fiscal austerity; they argue that the alternatives to neoliberalism are predicated on the spending of public money to create and facilitate a better society. Are there other ways for us – as humans – to

---

11   See Alcock, P. (2010), "Building the Big Society: a new policy environment for the third sector in England", *Voluntary Sector Review*, 1(3), 379–389.

12   For example, Macmillan Cancer Support contest the government's proposal to cut support to cancer patients by up to £94 a week go through as part of its controversial Welfare Reform Bill, debated in Parliament in October 2011. An additional proposal would see those cancer patients who require immediate financial help to cover extra costs following their diagnosis being forced to wait for six months instead of three to get the Personal Independence Payment (PIP), which replaces the Disability Living Allowance (DLA). See Macmillan.org.uk

13   The New Economics Foundation provides one such vision and bills its work as "economics as if people and the planet mattered." See www.newecomics.org

relate to each other and to the public sphere in which we see each other as neither cogs in a capitalist machine, nor as passive recipients of state-funded largesse, but as something else altogether? The multitude of possible responses to these questions, and the actions flowing from them, will shape our lives – as individuals and as societies – for generations to come.

Bibliography

Adams, M. (2002), *Interview at HEFCE*, Coventry, 27 March 2002.

Alcock, P. (2010), Building the Big Society: a new policy environment for the third sector in England, *Voluntary Sector Review*, 1(3), 379–89.

Altman, A. (1990), *Critical Legal Studies, A Liberal Critique*, Princeton University Press, Princeton, NJ.

Amar, A. (2000), *The Bill of Rights: Creation and Reconstruction*, Yale University Press, New Haven, CT and London.

Andrews, R. (2010), *Argumentation in Higher Education*, Routledge, London.

Archives (2005), *Comprehensive Spending Review, Section II: the Government's Key Objectives, Chapter 4: Efficient and Modern Public Services,* www.archive. official-documents.co.uk

Arendt, H. (1998), *The Human Condition*, University of Chicago Press, Chicago.

Arnold, K. (2002), Getting to the top; What role do elite colleges play?, *About Campus*, 7 (5), 4–12.

Bagguley, P., Mark-Lawson, J., Shapiro, D., Urry, J., Walby, S., and Warde, A. (1990), *Restructuring: Place, Class and Gender*, Sage, London.

Bailey, F. (1977), *Morality and Expediency; The Folklore of Academic Politics*, Basil Blackwell, Oxford.

Ball, S. (ed.) (1990), *Foucault and Education; Disciplines and Knowledge*, Routledge, London and New York.

Barnes, C. (1991), *Disabled People in Britain and Discrimination*, Hurst and Co., London.

Barnes, C. and Oliver, M. (1990), Disability rights: rhetoric and reality in the UK, *Disability and Society*, 10 (1), 111–16.

Barnes, C., Mercer, G., and Shakespeare, T. (1999), *Exploring Disability: A Sociological Introduction*, Polity Press, Cambridge.

Barnes, C., Oliver, M., and Barton, L. (eds) (2002), *Disability Studies Today*, Polity, Cambridge.

Barnett, R. (1990), *Idea of higher education*, Open University Press, Buckingham.

Barnett, R. (2000), *Realizing the university in an age of supercomplexity*, Open University Press.

Barrett, B. (1998), What is the function of a university?, *Quality Assurance in Education*, 6 (3), 145–51.

BBC News, 19 June 2003, *More rights for disabled workers*, news.bbc.co.uk

Bellamy, R. (2000), *Rethinking Liberalism*, Pinter, London and New York.

Bellow, G. and Minow, M. (1996), *Law Stories*, University of Michigan Press, Ann Arbor, MI.

Benjamin, S. (2002), *The Micropolitics of Inclusive Education: an Ethnography*, Open University Press, Buckingham.

Bhabha, H. (1994), *The location of culture*, Routledge, London.

Blair, T. 22 April 2005, Full text: Tony Blair's speech on asylum and immigration, *The Guardian*, www.guardian.co.uk

Birch, A. (1998), *The British System of Government*, Routledge, London and New York.

Bird, C. (1999), *The Myth of Individual Liberalism*, Cambridge University Press, Cambridge.

Blackburn, R. (ed.), (1993), *Rights of Citizenship*, Mansell, London.

Blake, N., Smith, R., and Standish, P. (1998), *The Universities we need: Higher Education after Dearing*, Kogan, London.

Blomley, N. (1994*), Law, Space and the Geographies of Power*, The Guilford Press, New York and London.

Blomley, N. (2000), *History, Geography and the Politics of Law*, Blackwell Publishers, Oxford and Malden, MA.

Blomley, N. and Clark, G. (1990), Law, Theory and Geography, *Urban Geography*, 11 (5), 433–46.

Blomley, N., Delaney, D. and Ford, R. (eds) (2001), *The Legal Geographies Reader*, Blackwell Publishers, Oxford and Malden, MA.

Borsay, A. (2004), *Disability and Social Policy in Britain since 1750: a History of Exclusion*, Palgrave, Basingstoke.

Brent, P. and Lane, A. (1972), *Godmen of India*, Penguin, London.

British Telecommunications plc (2002), Payphone use made easier for deaf people, *BT Today*, July 2002, p. 14.

Brooks, A. (1997), *Academic Women,* Open University Press, Buckingham.

Brueggeman, B. (1999) On (Almost) Passing, in *National Council of Teachers of English, Trends and Issues in Postsecondary English Studies*, University Of Illinois, Urbana, IL.

Bryman A. (2001), *Social research methods*, Oxford University Press, Oxford.

Burton D. (2000), *Research training for social scientists: a handbook for postgraduate researchers*, Sage, London.

Butler, R. and Parr, H. (eds) (1999), *Mind and Body Spaces, Geographies of Illness, Impairment and Disability*, Routledge, London and New York.

Byron University, (2003), Access consultants SENDA strategies report.

Campbell, J., and Oliver, M. (1996), *Disability Politics: understanding our past, changing our future*, Routledge, London and New York.

Canagarajah, A. (1999), Safe Houses in the Contact Zone: Coping Strategies of African-American Students in the Academy, in *National Council of Teachers of English, Trends and Issues in Postsecondary English Studies*, University Of Illinois, Urbana, IL.

Caprez, E. (ed.) (2002), *The Disabled Students' Guide to University*, Trotman, Richmond, Surrey.

Chouinard, V. (1999), Being Out of Place: Disabled Women's Explorations of Ableist Spaces, in Teather, E. (ed.), *Embodied Geographies: Spaces, Bodies and Rites of Passage,* Routledge, London and New York.

Clare, E. (1999), *Exile and Pride: Disability, Queerness and Liberation,* South End Press, Cambridge, MA.

Clare, E. (2001), Stolen bodies, reclaimed bodies: Disability and queerness, *Public Culture,* 13 (3), 359–65.

Clark, G. (2001), The Legitimacy of Judicial Decision-Making in the Context of Richmond v. Croson, in Blomley, N., Delaney, D. and Ford, R. (eds) (2001), *The Legal Geographies Reader,* Blackwell Publishers, Oxford and Malden, MA.

Clarke, J., Gewirtz, S., and Mclaughlin, E. (eds) (2000), *New Managerialism, New Welfare,* Sage, London.

Clarke, J. and Newman, J. (1997), *The Managerial State: Power, Politics and Ideology in The Remaking of Social Welfare,* Sage, London.

Cloke, P., Philo, C., and Sadler, D. (1991), *Approaching Human Geography: An Introduction to Contemporary Theoretical Debates,* Paul Chapman, London.

Cooper, J. and Vernon, S. (1996), *Disability and the Law,* Jessica Kingsley, London.

Cosgrave, P. (1978), *Margaret Thatcher; A Tory and her Party,* Hutchinson, London.

Cottell, C. 2 July 2005, Agencies to welcome disabled applicants, *The Guardian,* www.guardian.co.uk

Cresswell, T. (2004), *Place: A Short Introduction,* Blackwell, Malden, MA.

Cromwell Network, 1 February 2001, *Report on Cromwell's Work Culture.*

Curtis, P. 20 July 2005, Universities meetings access requirements, *The Guardian, www.guardian.co.uk*

Curtis, P. and MacLeod, D. 22 March 2005, Bursary Blues, *The Guardian, www.guardian.co.uk*

De Tocqueville, A. (1835) *Democracy in America, Volume I,* translated by Bowen, F. (1998), Wordsworth, London.

De Tocqueville, A. (1840) *Democracy in America, Volume II,* translated by Bowen, F. (1998), Wordsworth, London.

Dearlove, J. and Saunders, P. (2000), *Introduction to British Politics,* Polity Press, Cambridge.

Deem, R. (1998), New Managerialism in Higher Education – The Management of Performances and Cultures in Universities, *International Studies in the Sociology of Education* 8 (1), 47–70.

Deem, R. and Johnson (2000), Managerialism and university mananagers: building new academic communities or disrupting old ones?, in McNay, I. (ed.), *Higher education and its communities,* Open University Press, Buckingham.

Demerrit, L., and Lees, L. (2004), *Geographies of Collaborative Research: The Case of the ESRC CASE Studentship Programme,* unpublished.

Dent, J. (1993) (ed.), Utiliarianism, Everyman, London.

Department for Education and Employment (1999), *Disability Discrimination Act, An Introduction for Small and Medium-sized Businesses, Rights of Access to Goods, Facilities, Service and Premises*, Department for Education and Employment, London.

Department for Education and Employment (2001), *What Service Providers Need To Know*, Department for Education and Employment, London.

Department for Work and Pensions (2002), *DWP Research Report No. 169: Costs and benefits to service providers of making reasonable adjustments under Part III of the Disability Discrimination Act*, Stationery Office, London.

Department for Work and Pensions (2005), www.dwp.gov.uk

Disability Rights Commission (2001), *The Disability Discrimination Act 1995 (as amended by the Special Educational Needs and Disability Act 2001) Draft Code of Practice (Post 16) New duties (from 2002) in the provision of post-16 education and related services for disabled people and students*, Disability Rights Commission, Cheshire.

Disability Rights Commission (2002a), *Monitoring the DDA: is the law working?*, The Stationery Office, London.

Disability Rights Commission (2002b), *Disability Discrimination Act 1995, Code of Practice, Rights of Access, Goods, Facilities, Services and Premises*, The Stationery Office, London.

Disability Rights Commission (2002c), *Estates: the Disability Discrimination Act Part 4*, The Stationery Office, London.

Disability Rights Commission (2004), *Education for all: getting in, getting on, or getting nowhere?*, www.drc-gb.org

Disability Rights Commission (2005), www.drc-gb.org

Disability Rights Task Force (1999) *From Exclusion to Inclusion: A Report of the Disability Rights Task Force*, The Stationery Office, London.

Dominelli, L. (1997), *Anti-Racist Social Work*, Palgrave, Basingstoke and New York.

Dopson, S. and McNay, I. (1996) Organisational Culture, in Warner, D. and Palfreyman, D. (eds), *Higher Education Management: the key elements*, Open University Press, Buckingham.

Doyle, B. (1996), *Disability, Discrimination and Equal Opportunities: A Comparative Study of the Employment Rights of Disabled Persons*, Manseil, London and New York.

Doyle, B. (2000), *Disability Discrimination; Law and Practice*, Jordans, Bristol.

Drake, R. (1999), *Understanding Disability Policies*, Palgrave Macmillan, London.

Dreyfus, H. and Rabinow, P. (1983), *Michel Foucault: Beyond Structuralism and Hermeneutics*, Harvester Wheatsheaf, Hemel Hempstead.

Duberman, M. (2002), *Left Out: The Politics of Exclusion, Essays 1964–2002*, South End Press, Boston.

Durkheim, E. (1964), *The Division of Labour in Society*, The Free Press, New York.

Duryea, E. (2000), *The Academic Corporation; A history of college and university governing boards*, Falmer Press, New York and London.

Dwyer, C. and Bressey, C. (2008), *New Geographies of Race and Racism*, Ashgate, Aldershot.

Eagle, M. (2002), *Address to the Annual General Meeting of a Disability Organisation*, 25 November 2002, Southwark Cathedral, London. (Author's notes).

Esping-Andersen, G. (1990), *Three Worlds of Welfare Capitalism*, Princeton University Press, Princeton.

ESRC (2003), *Case Studentship Scheme*, www.esrc.ac.uk

Evans, E. (2004), *Thatcher and Thatcherism*, Routledge, London and New York.

Fairclough, N. (1989), *Language and Power*, Longman, Harlow.

Falzon, C. (1998), *Foucault and Social Dialogue,* Routledge, London.

Field Fisher Waterhouse (2003), *The Disability Discrimination Act 1995: How it affects owners, occupiers and managers of commercial property*, The European Legal Alliance, London.

Fincher and Jacobs, (eds) (1998), *Cities of Difference*, Guilford Press, New York.

Fitzpatrick, P. and Hunt, A. (eds) (1987), *Critical Legal Studies*, Basil Blackwell, Oxford.

Flyvbjerg, B. (2001), *Making Social Science Matter: Why social inquiry fails and how it can succeed again*, Cambridge University Press, Cambridge.

Forrest, J. (1998), *Namibia's Post-Apartheid Regional Institutions*, University of Rochester Press, Rochester, NY.

Foucault, M. (1972), *The Archaeology of Knowledge*, Tavistock, London.

Foucault, M. (1975), *Abnormal, in Two Lectures at the College de France, 1974-1975,* Marchetti, V., Salomoni, A. (eds), translated by Graham Burchell, 2003, Verso, London and New York.

Foucault, M. (1977), Intellectuals and power, in Bouchard, D. (ed.), *Language, Counter-Memory, Practice: Selected Essays and Interviews*, Cornell University Press, Ithaca, NY.

Foucault, M. (1979), *Discipline and Punish: the birth of the prison*, Penguin, Harmondsworth.

Foucault. M. (1980), *Power-knowledge: selected interviews and other writings,* edited by Gordon, C. Harvester Press, Brighton.

Freire, P. (1970), *Pedagogy of the Oppressed*, translated by Ramos, M (1998), Continuum, New York.

Freund, P. (2001), Bodies, Disability and Spaces: the social model and disabling spatial organisations, *Disability and Society*, 16 (5), 689–706.

Gane, M. and Johnson, T. (1993) *Foucault's new domains*, Routledge, London.

Gewirtz, P. (1997), *The Triumph of Antidiscrimination Law*, in Sarat, A. (ed.), *Race, Law and Culture, Reflections on Brown V. Board of Education*, Oxford University Press, Oxford and New York.

Gilroy, P. (1991), *'There Ain't no Black in the Union Jack': The Cultural Politics of Race and Nation,* University of Chicago Press, Chicago.

Gleeson, B. (1998), *Justice and the Disabling City*, in Fincher and Jacobs, (eds), *Cities of Difference*, Guilford Press, New York.

Gleeson, B. (1999), *Geographies of Disability*, Routledge, London.

Goldberg, D. (1993), *Racist Culture: Philosophy and Politics of Meaning*, Blackwell, Cambridge.

Goldberg, D. (1997), *Racial Subjects: writing on race in America*, Routledge/Taylor and Francis, London.

Gooding, C. (1994), *Disabling Laws, Enabling Acts; Disability Rights in Britain and America, Pluto Press*, Boulder, CO.

Gooding, C. (2000), *Disability Discrimination Act: from statute to practice, Critical Social Policy*, 20 (4), 533–51.

Gourley, B. (1999), *Against the Odds*, in Brennan, J., Huber, M., Shah, T., *What Kind of University? International Perspectives on Knowledge, Participation and Governance*, Open University Press, Buckhingham.

Greater London Authority (2002), *The Mayor's Annual Report*, Greater London Authority, London.

Gregory, D., Martin, R. and Smith, G. (1994), *Human geography: Society, space and social science*, Macmillan, Basingstoke.

Hahn, H. (2001) Adjudication or Empowerment: Contrasting Experiences with a Social Model of Disability, in Barton, L. (ed.), *Disability, Politics and the Struggle for Change,* David Fulton Publishers, London.

Hall, I., Clayton, P. and Johnson, P. (2005), *Arson and learning disability*, in Riding, T., Dale, C. and Swann, B. (eds), *The Handbook of Forensic Learning Disabilities*, Radcliffe, Milton Keynes.

Handley, P. (2001), *Analysis of the DDA 1992 (Australia),* unpublished.

Harvey, D. (1996), *Justice, Nature and the Politics of Difference*, Blackwell, Oxford.

Harvey, D. (2000), *Spaces of Hope*, Edinburgh University Press, Edinburgh.

Hebdige, D. (1988), *Subculture: The Meaning of Style*, Routledge, London.

Hekman, S. (2004) *Feminist Identity Politics: Transforming the Political*, in Taylor, D. and Vintges, K (eds), *Feminism and the Final Foucault*, University of Illinois Press, Urbana, IL.

Henkel, M. and Little, B. (eds) (1999), *Changing Relations between Higher Education and the State*, Jessica Kingsley, London and Philadelphia.

Hennessy, P. "Home Secretary: scrap the Human Rights Act", *The Telegraph*, 1 October 2011.

Henwood, K, and Pidgeon, N. (1993) *Qualitative Research and Psychological Theorising*, in Hammersley, M. (ed.), *Social Research: Philosophy, Politics and Practice*, Sage, London.

Higbee, J. (2001), Universal Design: Beyond Accommodations to Inclusion for Students with Disabilities, *paper presented at the Annual Conference of the American College Personnel Association*, March, Boston, MA.

Hillier, J. (1993), To boldly go where no planners have ever…, *Environment and Planning D: Society and Space*, Volume 11, pp. 89–113.

HMSO (1993), *Realising Our Potential: White Paper on policy issues in science and technology*, The Stationery Office, London.

HMSO (1995), *Disability Discrimination Act 1995*, The Stationery Office, London.

HMSO (1998), *Data Protection Act*, The Stationery Office, London.

HMSO (2001), *The Disability Discrimination Act 1995 as amended by the Special Educational Needs and Disability Act 2001,* The Stationery Office, London.

HMSO (2005), *The Disability Discrimination Act 2005*, The Stationery Office, London.

Hobbes, T. (1651), *Leviathan*, in Curley, E. (ed.) 1998, Hackett, Indianapolis.

Holloway, S. (2001), The Experience of Higher Education from The Perspective of Disabled Students, *Disability And Society*, 16 (4), 597-615.

hooks, b. (1982), *Ain't I A Woman; Black Women And Feminism*, Pluto Press, London.

hooks, b. (1992), *Black Looks: Race And Representation*, Turnaround, London.

hooks, b. (1994), *Teaching to Transgress: Education as the Practice of Freedom*, Routledge, New York.

hooks, b. (2000), *Feminist Theory: From Margin to Center*, South End Press, Cambridge MA -Links Experience To Theory.

hooks, b. (2000) Postmodern Blackness, *Postmodern Culture*, 1 (1) 1–15.

hooks, b. (2002), *Where We Stand: Class Matters*, Routledge, New York.

hooks, b. (2003) *Teaching Community: A Pedagogy Of Hope*, Routledge: New York, London.

HREOC (2003), Human Rights and Equal Opportunities Commission, Australia, www.hreoc.org.au

Hunt, A. and Wickham, G. (1994), *Foucault and Law: Towards a Sociology of Law as Governance*, Pluto Press, London.

Imrie, R. (2000), Disabling environments and the geography of access policies and practices in the United Kingdom, *Disability and Society*, *15(1),* 5–24.

Imrie, R. (2004a), Demystifying disability: a review of the ICF, *Sociology of Health and Illness*, 26 (3), 1–9.

Imrie, R. (2004b), Urban Geography, Relevance, and Resistance to the "Policy Turn", *Urban Geography*, 25 (8), 697–708.

Imrie, R. and Hall, P. (2001), *Inclusive Design: designing and developing accessible environments*, Spon Press/Taylor and Francis, London.

Imrie, R. and Thomas, H. (1997), Law, Legal Struggles and Urban Regeneration: Rethinking the Relationships, *Urban Studies*, 34 (9), 1401–8.

Jacobs, J. (1996), *Edge of Empire: Postcolonialism and the City*, Routlege, London and New York.

Kant, I. (1781), *Critique of Pure Reason*, translated by Meiklejohn, J. (1990), Prometheus, London.

Kant, I. (1785), *Grounding for the Metaphysics of Morals*, translated by Ellington, J. (1993), Hackett, London.

Kekes, J. (1997), *Against Liberalism*, Cornell University Press, Ithaca, NY and London.

Kells, H. (1992), *Self-Regulation in Higher Education; A Multi-National Perspective on Collaborative Systems of Quality Assurance and Control*, Jessica Kingsley, London and Philadelphia.

Kelly, J. (2003), Inspiration, *The Ragged Edge*, Jan/Feb 2003, 11–27.

Kitchin, R. (2000), *Disability, Space and Society,* Geographical Association, Sheffield.

Knopp, L. (1998), Sexuality and Urban Space: Gay Male Identity Politics in the United States, United Kingdom, and Australia, in Fincher and Jacobs, (eds), *Cities of Difference*, Guilford Press, New York.

Kogan, M. (1999), Academic and Administrative Interface, in Henkel And Little, *Changing Relationships Between Higher Education and The State*, Jessica Kingley, London.

Kogan, M. and Hanney, S. (2000), *Reforming Higher Education*, Jessica Kingley, London.

Komives, S. and Woodard, D. (1996), *Student Services: A Handbook for the Profession*, Jossey-Bass, San Francisco.

Konur, O. (2000), Creating Enforceable Civil Rights for Disabled Students in Higher Education: an institutional theory perspective, *Disability and Society*, 15(7), 1041–63.

Langton-Lockton, S. (2000), Language Matters, *Access By Design*, Issue 83, p. 14.

Lefebvre, H. (1991), *The Production of Space*, Blackwell, Oxford.

Lewis, C. (2002) Interview at Skill, London, 7 August 2002.

Linton, S. (1998), *Claiming Disability*, New York University Press, New York and London.

Lomas, L. (2000), *Senior Staff Members' Perceptions of Organisational Culture and Quality in HEI's in England* (Ph.D. thesis), University of Kent at Canterbury.

London Borough of Tower Hamlets (1997), *DDA 1995: Our Responsibilities (Handbook)*, Tower Hamlets Corporate Equalities Service, London.

Lucas, C. and Jones, J. (2011) "To recover from the riots we must rebalance the inequalities of society," *The Guardian*, 22 August 2011.

Lydon, J. (1995), *Rotten: No Irish, No Blacks, No Dogs*, Picador, London.

Macleod, D. 10 March 2003, Bristol rebuts bias claims by CRE head, *The Guardian, www.guardian.co.uk*

Maksidi, G. (1984), The *Rise of Colleges: Institutions of Learning in Islam and the West*, 1984, University of Edinburgh Press, Edinburgh.

Mansell, J. and Ericsson, K. (eds) (1996), *Deinstitutionalization and Community Living; Intellectual Disability Services in Britain, Scandinavia and the USA*, Chapman and Hall, London.

Marshall, J. (1990) Foucault and educational research, in Ball, S. (ed.), *Foucault and Education; Disciplines and Knowledge*, Routledge, London and New York.

Martin, L. and Gutman, H (eds) (1988), *Technologies of the Self: a Seminar with Michel Foucault*, Tavistock: London.

Marx, K. (1992), *Capital: A Critique of Political Economy – Volume One*, translated by Fowkes, B., Penguin, London.

Maslin, K. (2001), *An introduction to the philosophy of mind*, Polity, Cambridge.

Massey, D. (1994), *Space, Place and Gender*, University of Minnesota Press, Minneapolis.

Massey, D. (1997), *A Global Sense of Place*, in Barnes, T. and Gregory, D., eds, *Reading Human Geography*, Arnold, London.

Massey, D., Allen, J. and Sarre, P. (eds) (1999), *Human Geography Today*, Polity Press, Cambridge.

Matsuda, M., Lawrence III, C, Delgado, R., and Crenshaw, K. (1993), *Words that wound: critical race theory, assaultive speech, and the First Amendment*, Westview Press, Boulder, San Francisco, Oxford.

May, J., Cloke, P. and Johnsen, S. (2005) Re-phasing Neoliberalism: New Labour and Britain's Crisis of Street Homelessness, *Antipode* 37 (4), 703–30.

Meager, N., Doyle, B., Evans, C., Kersley, B., Williams, M., O'Regan, S. and Tackey, N. (2002), *Monitoring the Disability Discrimination Act 1995: DfEE Research Report RR119*, Department for Education and Employment, London.

Mercer, G. (2002), Emancipatory disability research, in Barnes, C., Oliver, M., and Barton, L., (eds), *Disability Studies Today*, Polity, Cambridge.

Merriam-Webster (2002), *Merriam-Webster's Dictionary*, London.

Mill, J. (1859), *On Liberty*, Cambridge University Press, Cambridge.

Miller, T. (1992*), The Second Fifty Years: Promoting Health and Preventing Disability*, (Institute of Medicine), Washington DC: National Academies Press.

Minow, M. (1991), *Making all the Difference: Inclusion, Exclusion and American Law*, Cornell University Press, Ithaca, NY.

Mitchell, D. (2000), *Cultural Geography: A Critical Introduction*, Blackwell, Malden, MA.

Monaghan, K. (2011), "The Equality Act is one. Will the coalition's birthday gift be to repeal key provisions?", *The Guardian*, 3 October 2011.

Monbiot, G. (2000), *Captive State: The Corporate Takeover of Britain*, Macmillan, London.

Moran, M. (2003), *Rethinking the reasonable person: an egalitarian reconstruction of the objective standard*, Oxford University Press, Oxford.

Morley, L. (2004), *Theorising Quality in Higher Education*, Institute Of Education, London.

Morris, J. (1991), *Pride Against Prejudice; Transforming Attitudes to Disability*, Women's Press, London.

Morris, J. (1996) (ed.), *Encounters with Strangers: Feminism and Disability*, The Women's Press, London.

Muir, H. (2009), Trevor Phillips: a career in crisis, *The Guardian*, 28 July 2009.

Mulholland, H. 25 May 2005, Prejudice 'still blocking disabled people's path to work', *The Guardian*, www.guardian.co.uk

Nash, J. and Calonico, J. (1993), *Institutions in Modern Society: Meanings, Forms and Character*, Rowman and Littlefields, Oxford.

National Board of Employment, Education and Training Australia (1994), *Guidelines for Disability Services in Higher Education, Commissioned Report No. 29*, DEET, Canberra.

New York Times (2005), *College Times*, www.nytimes.com

Newman, F., Couturier, L., and Scurry, J. (2004). *The Future of Higher Education: Rhetoric, Reality, and the Risks of the Market.* San Francisco: Jossey Bass.

Newman, J. (2001), *Modernising Governance: New Labour, Policy and Society*, Sage, London.

Newman, J. (2000), Beyond The New Public Management, in Clarke, J., Gewirtz, S., Mclaughlin, E. (eds), *New Managerialism, New Welfare*, Sage, London.

ODPM (2002), *Planning and access for disabled people: a good practice guide*, www.odpm.gov.uk

Oliver, M. (1990), *The Politics of Disablement*, Macmillan, Basingstoke.

Oliver, M. and Barnes, C., (1998), *Disabled People and Social Policy: From Exclusion to Inclusion*, Longman, London.

Olssen, M. (1999), *Michel Foucault: materialism and education*, Bergin and Harvey, Westport, CT.

Olson, G. (1994), bell hooks and the Politics of Literacy, *JAC, Journal of Composition Theory*, 14 (1), 1–26.

Palfreyman, D. and Warner, D. (eds), (2002), *Higher Education Law*, Jordans, Bristol.

Pannick, D. (1987), *Judgment of the Judicial Committee of The Privy Council*, www.privy-council.org.uk

Phillips, A. (2004) Defending Equality of Outcome, *Journal of Political Philosophy*, 12 (1), 1–19.

Philo, C., Parr, H., and Burns, N. (2002), *Social differences: locals, incomers, gender, age and ethnicity* (draft), Findings Paper No.8, Department of Geography and Topographic Science, University of Glasgow, www.geog.gla.ac.uk/olpapers/cphilo009.pdf

Power, M. (1999), *Audit Society: Rituals of Verification*, Clarendon Press, Oxford.

Prentis, D. (2005), *Comprehensive Spending Review*, www.unison.org.uk

Pue, W. (1990), Wrestling with law: (geographical) specificity versus (legal) abstraction, *Urban Geography*, 11, 221–35.

Q, K. (2011), "Judgment Day", *The Economist*, 13 October 2011.

Race, D. (2002) (ed.) *Learning Disability: A Social Approach*, Routledge, London.

RADAR (2003), This time let's hope the bill makes the Queen's speech, *RADAR Bulletin*, March 2003, p. 3.

Radcliffe, S. (1999), *Popular and State Discourses of Power*, in Massey, D., and Allen, J., (eds), *Human Geography Today*, Polity Press, Cambridge.

Ramsay, M. (1997), *What's Wrong With Liberalism? A Radical Critique of Liberal Political Philosophy*, Leicester University Press, London and Washington, DC.

Razack, S. (2002), *Race, Space And The Law: Unmapping A White Settler Society*, Between The Lines, Toronto.

Razack, S. (1998), *Looking White People in the Eye: Gender, Race and Culture in Courtrooms and Classrooms*, University of Toronto Press, Toronto and Buffalo, NY.

Razack, S. (1991), *Canadian Feminism and The Law*, Second Story Press, Toronto.

Read, F. and Read, E. (2008), Learning Disability and Serious Crime – Arson, *Mental Health and Learning Disabilities Research and Practice*, 2008 (5), 210–23.

Reinders, H. (2000), *The Future of the Disabled in Liberal Society; an Ethical Analysis*, University of Notre Dame Press, Notre Dame, IN.

Ritchie, J. and Boardman, K. (1999), Feminism In Composistion; Inclusion, Metonymy And Disruption in National Council of Teachers of English, *Trends And Issues In Postsecondary English Studies*, University of Illinois, Urbana, IL.

Rose, N. (1987), Beyond the Public/Private Division: Law, Power and the Family, in Fiztpatrick P., and Hunt, A. (eds), *Critical Legal Studies*, Basil Blackwell, London.

Roulstone, A. (2003), The Legal Road to Rights? Disabling Premises, Obiter Dicta and the Disability Discrimination Act 1995, *Disability and Society*, 18 (2), 117–31.

Russell, B. (2000), *Happiness*, Four Seasons, Kingston upon Thames.

Said, E. (1978), *Orientalism: Western Conceptions of the Orient*, Penguin: London

Salant, P. and Dillman, D. (1994), *How to conduct your own survey*, John Wiley and Sons, New York.

Sarat, A. (1987) (ed.), *Race, Law and Culture, Reflections on Brown V. Board of Education*, Oxford University Press, Oxford and New York.

Savage, M., Barlow, J., Dickens, P., and Fielding, T. (1992), *Property, Bureaucracy and Culture*, Routledge, London and New York.

Sayer, A. (1999), *Realism and Social Science*, Sage, London.

Schmitt, C. (1976), *The Concept of the Political*, translated by Schwab, G., Rutgers University Press, New Brunswick, NJ.

Schuck, P. (2000), *The Limits of Law; Essays on Democratic Governance*, Westview Press, Oxford and Boulder, CO.

Schumaker, J. and Carr, S. (eds) (1997), *Motivation and Culture*, Routledge: London and New York

Scott, P. (2000), *University Leadership: the Role of the Chief Executive,* Open University Press, Buckingham.

Shakespeare, T., Gillespie-Sells, K., and Davies, D. (1996), *The Sexual Politics of Disability,* Continuum, London.

Stiker, Henri-Jacques, (2000), *A History of Disability*, translated by Sayers, W., University of Michigan Press, Ann Arbor, MI.

Stone, D. (1985), *The Disabled State*, Macmillan, London.

Struck, D. 3 May 1997, Clinton Dedicates Memorial, Urges Americans to Emulate FDR, *The Washington Post*, www.washingtonpost.com

Stubbs, M. (1983), *Discourse Analysis: The Sociolinguistic Analysis of Natural Language,* Basil Blackwell, Oxford.

Sullivan, P. (1996), *Days of Hope: Race and Democracy in the New Deal Era*, University of North Carolina Press, Chapel Hill, NC.

Swain, J., French, S., Barnes, C., and Thomas, P. (eds), (2004), *Disabling Barriers, Enabling Environments*, Sage, London.

Tawney, R. (1998), *Secondary Education for All: A Policy for Labour*, Continuum, London.

Taylor, M. 10 August 2005, Top-up fees will deter students, survey reveals, *The Guardian,* www.guardian.co.uk

Taylor, R., Steele, T. and Barr, J. (2002), *For a Radical Higher Education, After Postmodernism,* Open University Press, Buckingham.

THES (2005), *Times Higher Education Supplement,* www.thesis.co.uk

Toynbee, P. (2009), "Harman's law is Labour's biggest idea for 11 years," *The Guardian,* 13 January 2009.

Transport for London (2001), *The Mayor's Transport Strategy: Executive Summary and Accessibility Action Plan,* Greater London Authority, London.

Tuan, Y. (1977), *Space and Place: The Perspective of Experience,* University of Minnesota Press, Minneapolis.

Tuckman, B. (1999), *Conducting educational research,* Harcourt Brace, Fort Worth.

UNESCO (2000), *Inclusive Education,* portal.unesco.org/education

U.S. DOJ, (1992) *ADA/ABA Handbook: Accessibility Guidelines for Buildings and Facilities,* Washington, D.C.

Waldron, J. (1993), *Liberal Rights: Collected Papers 1981-1991,* Cambridge University Press, Cambridge.

Wapshott, N. and Brock, G. (1983), *Thatcher,* MacDonald, London and Sydney.

Ward, I. (1998), *An Introduction to Critical Legal Theory,* Cavendish, London.

Wickham, G. and Pavlich, G. (eds) (2001), *Rethinking Law, Society And Governance; Foucault's Bequest,* Hart Publishing, Oxford.

Williams, J. (ed.) (1997a), *Negotiating Access to Higher Education: The Discourse of Selectivity and Equity,* Open University Press, Buckingham.

Williams, P. (1997b), *Alchemy of Race and Rights,* Harvard University Press, Cambridge, MA.

Wintour, P. 18 May 2005, Controversy over incapacity benefit reform, *The Guardian,* www.guardian.co.uk

Wodak, R. (ed.) (1989), *Language Power and Ideology: Studies in Political Discourse,* Benjamins, London.

Woodhams, C. and Corby, S. (2003), Defining Disability in Theory and Practice: A Critique of the Disability Discrimination Act 1995, *Journal of Social Policy,* 32, 2. pp. 159–78.

Young, I. (1994), *Justice and the Politics of Difference,* Princeton University Press, Princeton, NJ.

Younge, G. 5 September 2005a, Left to sink or swim: Tragic events in New Orleans have laid bare America's bigotry and exposed the lie of equal opportunity, *The Guardian,* www.guardian.co.uk

Younge, G. 25 July 2005b, No tails or tridents, *The Guardian,* www.guardian.co.uk

Younge, G. 16 May 2005c, Detox our racist culture, *The Guardian,* www.guardian.co.uk

Zifcak, S. (1994), *New Managerialism: Administrative Reform in Whitehall and Canberra,* Open University Press, Buckingham.

Index